Hydraulics
Field Manual

Other McGraw-Hill Books of Interest

Hydraulics
Field Manual

Robert O. Parmley, P.E.
Editor-in-Chief

McGraw-Hill, Inc.

New York St. Louis San Francisco Auckland Bogotá
Caracas Lisbon London Madrid Mexico Milan
Montreal New Delhi Paris San Juan São Paulo
Singapore Sydney Tokyo Toronto

Library of Congress Cataloging-in-Publication Data

Parmley, Robert O.
    Hydraulics field manual / Robert O. Parmley.
        p.   cm.
    Includes index.
    ISBN 0-07-048556-9
    1. Hydraulics — Handbooks, manuals, etc.   2. Hydraulic engineering —
Handbooks, manuals, etc.   I. Title.
TC160.P34   1992
627 — dc20   92-12036
                                                            CIP

ISBN   0-07-048556-9

*The sponsoring editor for this book was Harold B. Crawford, the
editing supervisor was Stephen M. Smith, and the production supervisor was Pamela A. Pelton. It was set in Century Schoolbook by
Progressive Typographers.*

Printed and bound by R. R. Donnelley & Sons Company.

This book is dedicated to the memory of those unsung engineers who have contributed to the mastery and civilized usage of water.

Contents

Section 4. Pumps   59

Section 5. Weirs, Flumes, and Orifices   115

Section 6. Flow in Pipes   145

Section 8. Storage and Fire Protection   243

Section 9. Estimating Flows in the Field   267

Preface

Water, the most common liquid in this world, covers more than 70 percent of this planet's surface. It creates oceans, produces lakes, makes rivers, fills voids in the earth, falls in the form of precipitation, provides moisture in the air, and is frozen in the polar ice caps.

The continuous activity of this substance shapes and alters the earth's surface. Floods, erosion, ocean wave action, precipitation, glacial activity, and snow-melt and freeze-thaw cycles are some of the more dramatic forces which mold the facial features of the terrestrial landscape.

Humankind's earliest recordings reflected the significant effect and importance that water played in their daily lives. Civilizations either flourished or perished, relative to the availability of water. In essence, water is such a basic component that life, as we know it, could not exist without it. For this reason, some individuals throughout history attempted to accurately record, study, and understand the physical activities of water so that they could improve the quality of human existence.

Egyptian inscriptions as early as 3000 B.C. describe floods of the Nile River, and precise floodmarks recorded as early as 1800 B.C. have been discovered. During that period, the Nilometer or Egyptian staff gauge was developed. Historians tell of rainfall measurements being taken in India in the fourth century B.C. The Old and New Testaments contain over 1500 direct references to water and related items such as rain, dew, wells, springs, and seas.

Homer, Plato, and Aristotle attempted to explain rainfall and stream flow with elaborate, but unsound, hypotheses which apparently were not questioned with any authority until the fifteenth and sixteenth centuries A.D. Perhaps it was considered a form of heresy to challenge the ideas of the early classic philosophers. In any event, practical men like da Vinci, Palissy, Perreault, Mariotte, and Halley broke the chains of darkness by using logic and keen observation to explain some of the mysteries, so long hidden, concerning the hydrologic cycle and related effects.

The works of Bernoulli, Venturi, de Chézy, and others of like mind paved the way for rapid enlightenment in the study of hydraulics. Structures such as bridges, canals, dams, and dikes soon were being designed by engineers with a much-improved knowledge of fluid mechanics. However, little was known at that time about probable volumes of water to which these structures could be exposed. Understanding of this area of hydrology came of age much later and is still being refined with the aid of modern computers and a greater range of recording stations. However, bear in mind that the National Weather Service states that the average density of official rain gauges in the United States is only about 1 per 250 square miles, which results in relatively scattered sampling over a vast area.

Many ancient civilizations built waterways, aqueducts, canals, irrigation systems, and dams, but the Romans, at the height of their power, constructed some superior facilities throughout their known world. According to the Athens News Agency in August 1990, ancient Roman water reservoirs discovered beneath streets and buildings were recently used for landscape irrigation and other nonpotable uses to relieve Athens' current water shortage. It is believed that these reservoirs date back to the reign of the Roman Emperor Hadrian, who ruled the Empire between A.D. 117 and 138. These subsurface masonry reservoirs were supplied by aqueducts delivering water from the adjacent Pendeli Mountains.

While transporting water through artificially constructed conduits was employed on a limited basis from early times, this method of conveyance did not fully sprout until the late 1800s, when people became more skilled in fabricating pipe. Around the turn of this century, the modern understanding of public health and community sanitation dictated the need for public sewers and safe water distribution systems. With this demand, engineers needed a better understanding of the hydraulics of piping networks. A vast number of individuals contributed to our present knowledge of network piping hydraulics, and many will be discussed in the appropriate sections of this manual. However, no one, in my opinion, contributed more to the fundamental knowledge of modern-day hydraulics than the celebrated Irish engineer Robert Manning. According to ISCO's *Open Channel Flow Measurement Handbook,* a version of his formula was presented in 1889 at the meeting of the Institute of Civil Engineers of Ireland. A later modification was recommended for international use in 1936 by the Third World Power Conference. Because of its simple form and generally good results, Manning's formula has become the most widely used of all gravity flow formulas for open-channel flow calculations.

Humankind's continuous connection with water and the long struggle to master its physical activities would require a lengthy series of books to adequately tell the story. This manual will only lightly touch some of these areas to set the stage; its main purpose is to provide a basic outline of hydraulics and fundamental data to technically support "field personnel" on reconnaissance missions involving the control of water.

In developing this volume, the engineering literature and related technical publications were systematically searched. Various technical societies and institutions were contacted for assistance. Manufacturers' manuals for measuring and recording equipment were consulted. Following an exhaustive assembly of material, the data was sorted and culled into general categories. With practicality in mind, each section of material was boiled down to basics and then edited into the format published herein.

The user must be aware of the fact that the full range of hydraulics is gigantic and certainly cannot be housed in one publication. Therefore, it became an editorial judgment call to limit the contents to what I believe is a primary database for engineering field personnel. Since it is impossible to tailor this manual to each individual's specific needs, one or two blank pages have been added to the end of each section for the user's notes and data, to personalize his or her copy.

Much of the material and data contained in this manual were

obtained from a wide range of sources. Engineering handbooks, encyclopedias, technical publications, manufacturers' literature, and various individuals supplied valuable data and granted permission for use in this work. Proper credit and recognition has been placed on the page or pages where it is located. In summary, I thank these sources and salute the unsung contributors who are generally unknown to the public and often forever remain in the shadow of their projects.

Additionally, I want to thank Lana and Ethne for typing the manuscript and related correspondence. Special recognition for Wayne's photography and graphic design assistance is warranted; the images at the beginning of each section were prepared under his direction.

A final note: Realizing the important and expanding role of women in all engineering fields, every effort has been made in this manual to present material in gender-neutral language. However, for certain terms, such as manhole, no generally accepted gender-neutral equivalents exist, and occasionally, to avoid the awkwardness of the "he/she," "her/his," or "person" types of grammatical restructuring, the older usage of the words "he," "his," and "man" to refer to people in general has been allowed to stand. In all cases, the words should be taken in a purely generic sense, intended to apply to women and men equally.

Robert O. Parmley, P.E.

Section

1

Hydrology

Hydrology is the scientific study of water, especially in its natural occurrence, detailing its typical characteristics, control, distribution on land, precipitation, storage, runoff, evaporation, and ultimate return to atmosphere and the ocean. Since a basic understanding of this science is necessary to fully grasp field hydraulics, this section will present a brief overview of hydrology and its related areas in an attempt to set the stage for this manual.

1-1 Water

Water is the most common natural substance on this planet. It covers almost three-quarters of the earth's surface, filling the oceans, seas, lakes, rivers, creeks, streams, swamps, and much of the soil. Moisture is in the air we breathe and the clouds in the sky.

Second only to air, water is the single most essential component necessary in nature to sustain life. Without water, life as we know it could not exist. Every living thing, be it plant, animal, or human being, must have water to live. Every living thing consists mostly of water.

It has been estimated that there are about 326 million cubic miles of water on earth. The world's supply of water is dispersed as follows: 97 percent in the oceans and seas, 2 percent in polar ice caps and glaciers, 0.6 percent in groundwater, 0.3 percent in the atmosphere, and 0.1 percent in lakes, inland seas, salt lakes, rivers, streams, and wetlands. Therefore, about 3 percent of the earth's water is fresh.

Water is the only substance naturally present in three different states within the earth's normal temperature range: a liquid, a solid (ice), and a gas (water vapor). It is the only natural substance that physically expands when it is heated and when it freezes. For this fact, ice floats on liquid water. If water contracted when it froze, any volume of ice would weigh more than an equal volume of water and therefore sink. If ice sank, each winter more and more ice would accumulate on the bottoms of oceans, lakes, and rivers. Summer heat could not reach these ever-growing ice packs to melt them, and the cycle would continue until all the earth's water would be locked in solid ice, except for a very thin stratum of water over the ice during summer months. Life as we know it would certainly end or never have begun in the first place.

Water is formed by two flammable gases (hydrogen and oxygen) into a substance that is impossible to ignite. Water is held together by two forces, chemical bond and hydrogen bond.

Chemical bonds are forces that hold two hydrogen atoms and one oxygen atom together, thus forming a water molecule. Each hydrogen atom has one electron in orbit circling its nucleus with each of these atoms having room for two additional electrons. The oxygen atom has six electrons in its outer orbit, but room for eight. The result is that the hydrogen and oxygen atoms each fill their respective empty areas by cross-sharing their electrons.

Hydrogen bonds are forces that unite water molecules. Water molecules have a bulge or lopsided form because the two hydrogen atoms balloon from one end of the oxygen atom. The hydrogen end of the water molecule is positively charged while the opposite end of the molecule is negative, thus linking together because of opposite electrical charges attracting each other.

1-2 Water Effects

Beyond the simple fact that life could not exist on earth without water, this substance continually affects our planet and lives in innumerable areas. Rain, hail, snow melt, floods, and erosion constantly change the surface appearance. Wetlands, rain forests, and deserts sprinkle the globe. Water is unevenly dispersed throughout the world, and its distribution dictates population concentrations and the rise and fall of civilizations.

Ancient plagues sometimes were the result of water-borne contagious diseases that altered cultural behavior and decreased populations. Catastrophic floods place humankind at their mercy. Climatic conditions with accompanying humidity, precipitation, and lack thereof, have forced people to modify their environment to seek reasonable protection from the natural ravages of the ongoing hydrologic cycle.

Water and its many effects are so interwoven with our physical existence that our recorded history reflects it in more ways than we are consciously aware. The Bible makes well over 1500 direct references to water and its correlation with wells, springs, seas, rain, dams, channels, and, of course, the great flood that Noah and his family survived. In direct contrast to water, the lack of an adequate supply of this substance causes arid conditions, and in extreme cases, deserts result.

We cannot escape the basic fact that water is central to the history of humankind. Civilizations have flourished or vanished as a direct effect of their respective water supply. Only recently have we come to realize and understand that the steady growth of industry and population places an ever-increasing burden on our fixed supply of water. We have learned by costly experience how easily the water resources can become polluted. In order to continue our technology advance and supply the population with potable water, we must manage the freshwater supply with constant vigilence to safeguard this vital substance which has been taken for granted for so long.

1-3 Hydrologic Cycle

The waters of this planet are in constant movement from oceans, to atmosphere, to land, and thence back to the oceans. This never-ending natural process is called the *water cycle* or, more commonly, the *hydrologic cycle.*

The sun's radiation produces heat and evaporates surface water from the oceans, causing an invisible water vapor to rise and ultimately forming clouds as it cools. Further cloud cooling yields precipitation in the form of rain, sleet, hail, and snow which falls largely on the ocean. The remainder of this moisture is deposited on the land. Some percolates or infiltrates into the soil and thence into the groundwater. The balance flows into the drainage systems — streams, creeks, rivers, wetlands, and lakes — and finally empties back into the oceans. Only that precipitation which falls on the polar ice caps and high mountain peaks is side-tracked in the cycle. Refer to Fig. 1-1 for a basic schematic of the hydrologic cycle. The illustration is necessarily oversimplified to allow graphic support to this brief description.

The earth has a tremendous quantity of water, estimated

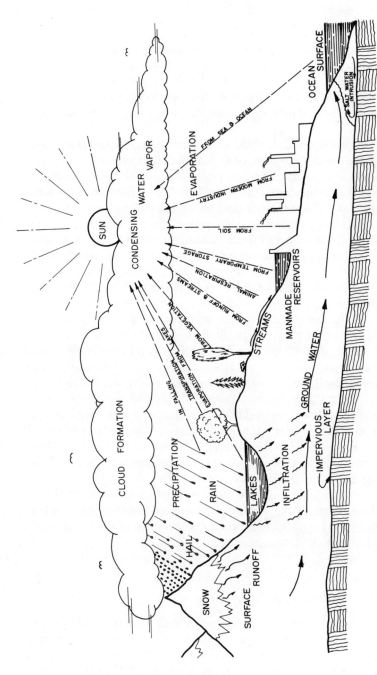

FIGURE 1-1 Hydrologic cycle.

5

to total approximately 326 million cubic miles, as stated previously. However, there is as much water on earth today as there ever was, since the water volume does not change. The water you drink today may well have quenched the thirst of King David or perhaps Roman slaves building an ancient aqueduct.

At some time, all of the earth's water enters the atmosphere in the form of water vapor as a result of the evaporation process. Clouds form and, as cooling occurs, precipitation is produced, yet the atmosphere holds only 1/1000 of 1 percent of the total quantity of water, according to leading authorities.

Following the fall of precipitation, the cycle continues in the form of runoff and ultimate drainage into the final ocean reservoir. It must be stated that all phases of the cycle are occurring simultaneously on a worldwide basis, but studied in terms of a limited site, the volumes vary extremely: from floods to droughts.

The absence of practical investigation and limited knowledge of the physical world resulted in philosophic speculation of hydrology problems by such historic figures as Homer, Plato, and Aristotle. These untested theories and assumptions persisted until the fifteenth century A.D. when practical men such as Palissy and daVinci described the hydrologic cycle very closely to the present-day understanding of this phenomenon. Pierce Perreault computed measurements of rainfall and runoff for France's Seine River in the 1700s which supplied quantitative proof that runoff volume was only about one-sixth the rainfall. This finding was supported by data collected by Mariotte and later reinforced by the English astronomer Halley demonstrating that moisture evaporated from the seas was sufficient to supply stream flow, thus establishing a practical analysis of the hydrologic cycle.

1-4   Hydrology in Engineering

Civil engineers use the science of hydrology generally in connection with the planning, design, and operation of hydraulic structures and systems. Estimates of flood flows and storm events are used in design of highways, bridges, and dams. Reservoir sizing for waterworks and shoreline structure design utilize hydrology in planning these facilities. Floodplain boundaries are computed to anticipate potential areas affected by excessive storm waters. Knowledge of the fundamentals of hydrology is an essential segment of the civil engineer's knowledge.

Generally, hydrology problems are individual or unique and deal with a site-specific location having a distinct set of physical conditions. These problems involve estimates of extremes which rarely are observed and typically have no historical data available. Quantitative conclusions of one analysis may not necessarily be transferable directly to another situation. Therefore, it is extremely important to collect accurate data and use proper methods of intepretation and sound judgment to reach a reasonable conclusion.

The central core of hydrology as a science is the measuring of each phase of the terrestrial segment of the hydrologic cycle, plus understanding the various physical processes of water. The practical concern is not with the cycle itself, but with its variations. Since precipitation certainly does not occur evenly throughout the globe, there are normally wet and dry locations. At any point in time, there are significant variations from season to season, year to year, and critical event to catastrophic occurrences that must be considered.

Hydrologists measure present variations in the hydrologic cycle and interpret this data to predict future variations or peaks in this natural cycle. Historical records are the best predictor for anticipating future events and charting forecasts. The history of precipitation in the continental United States is accumulated from the records of many thousands of gauges. The present network has over 13,000 precipitation gauges measuring rain and snowfall. About 3000 of these recording facilities weigh precipitation and tally the accumulated moisture depth, time, and duration, thus showing the intensities of each precipitation event. All hydroelectric dams and many flood-control structures have recording staff gauges to continuously chart water levels and spring melt and flood conditions.

It should be mentioned that fog and dew consist of water droplets so small that their falling velocities are negligible. The process of "fog drip" is an important source of water for vegetation, especially in rainless summer on the Pacific Coast. However, this precipitation cannot be measured.

Meteorologists, oceanographers, and hydrologists are learning more about the water cycle and its influence on humanity through new technology using earth satellites, high-speed telemetry, modern computers, and radar. These tools are improving our ability to prevent and/or minimize natural weather hazards, as well as planning that future generations do not receive a network of polluted rivers or restricted freshwater supply.

Refer to Sec. 2 for further discussion and an expansion of hydrology and hydraulics as it relates to the practice of engineering.

References

H. E. Babbitt et al., *Water Supply Engineering,* 6th ed., McGraw-Hill, 1962.
R. K. Linsley, *Water-Resources Engineering,* 2d ed., McGraw-Hill, 1972.
R. K. Linsley et al., *Hydrology for Engineers,* 3d ed., McGraw-Hill, 1982.
J. C. Stevens, *Water Resources Data Book,* 3d ed., Leupold & Stevens, 1978.
World Book Encyclopedia, 1974 ed., "Water" and related articles.

NOTES

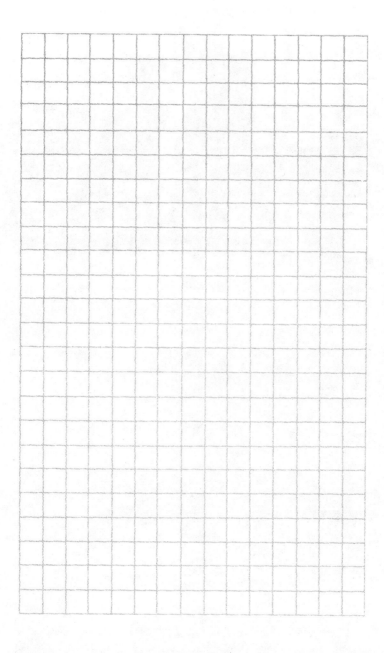

Hydraulics

2-1   Hydraulic Engineering

Water hydraulic engineering is defined as that branch of civil engineering concerned with the study, planning, design, erection, construction, and operation of waterworks, sewerage treatment facilities, dams, wells, drainage systems, water-powered generating plants, flood control structures, fire protection systems, irrigation networks, and similar facilities.

Hydraulics is the science of controlling and utilizing liquid pressure, specifically, the application of hydrodynamics and fluid mechanics to solve engineering problems. The term *hydraulic* literally means to be operated or be affected by the action of water or other fluids of low viscosity.

Hydraulic engineering is not a recent discipline. The engineers of old practiced this science under extremely limited conditions and without the benefit of the data and tools available today. Their basic equipment was good judgment and utilizing the "trial and error" approach. Fortunately, some of their past achievements were recorded. As enlightenment developed, piece by piece, we now stand with a vast reservoir of experience and acquired data, making our modern projects more reliable and cost-effective.

2-2   Historical Record

The history of civilized urbanization is, in a true sense, the record of how man has used water. His management of water directly affected the ability to live in close contact with his fellow men, without endangering their health, comfort, safety, and general quality of life. Community cleanliness, from potable water to the disposal of natural accumulation of the wastes occurring from living processes, appears to be the key factor for city survival, not to mention the quality of life.

It is attributed to Alexander the Great that no city in excess of 100,000 inhabitants could long survive; that "person to person" contacts would result in human destruction by man's desire to live in close proximity with his neighbors. It was disease and pollution, not hatred and ultimate war, that was the culprit. It was the biblical Mosaic law that established the foundation of hygiene and sanitation.

Engineering historians have had limited success in obtaining a clear record of unbroken documentation of man's water usage from early days to the present time. Searching the records reveals only fragments of data from early times, but certainly enough to disclose that sewer systems, wells, dams, canals, reservoirs, and related facilities played a major role in man's civilized progress.

The lack of precise historical records of man's management of water is unfortunate. However, it is this writer's opinion that, if such records could be discovered, civilization's progress would be directly proportional to humankind's usage of water. Success of individual cities could certainly be measured by their respective abilities to provide the population with pure drinking water, proper disposal of municipal sewage, and protection of the area from storm and floodwaters.

On the other hand, perhaps, the gaps in the historical records speak for themselves and are revealing proof of the rise and fall of communities. Significantly, the gaps in sanitary sewer records are verifications that the use of sewers for the purposes of disposing of sanitary wastes were not continuous throughout history. Comparing these periods with parallel records gives credence to the fact that if sanitary engineering principles are not followed, man cannot live safely in an environment of his own physiological waste products.

A prime example of this gap was recorded many times in the Dark Middle Ages. Public health was generally nonexistent. Man had lost the bright light of civilization established in earlier times and actually impeded or severed his development of sanitation methods and social progress.

Cities, towns, and villages were overrun with filth. Vermin infiltrated every area, from decaying debris piles to water sources. Food storage was always endangered and waste disposal was crude. The "window method" of disposal of wastewater and garbage was common practice. Periodic epidemics roared through unsuspecting communities, leaving the population puzzled by the resulting death and destruction. Superstitions replaced sanitation. The connection between disease transmission and water as a common carrier was not yet understood.

The Dark Middle Ages, when human social progress was severely retarded, was an "unlit" period of filth and neglect. The fact that these centuries were subsequent to an earlier historical period when human beings practiced personal hygiene and community cleanliness makes the "darkness" age almost inexcusable.

Early records of sewers, wells, dams, and waterways are intertwined with historical communities such as Babylon and Ninevah in the seventh century B.C., Assyrian cities, Egypt's sanitary canals, Carthage, Alexandria, Rome's aqueducts and its Cloaca Maxima sewer system, sewers of ancient Crete, Jerusalem's drains of biblical times, the sewers of Greece during her period of greatness, dams on the Tigris River, and irrigation networks in the Nile Valley.

It is well documented that Frontius, Rome's water commissioner, proclaimed in an edict that it is "necessary that a part of the supply flowing from the watercastles shall be utilized not only for cleaning our city, but also for flushing sewers."

The earliest pipes were natural tubes, perhaps hollow logs or bamboo. The Greeks used clay pipes to transfer water as early as 500 B.C. The Romans used lead pipes to transport water from their masonry aqueducts to various parts of the city. Copper water pipes date back to the Egyptian pyramids. It is recorded that Julius Caesar designed a tamped-concrete-pipe water distribution network for the city of Cologne, Germany, parts of which were used for over 1800 years. Brick sewers were also used extensively from times immemorial. Cast-iron pipe was first used in the water network in Versailles, France in 1664.

During the Industrial Revolution, many waterways were damned to form mill ponds for hydropower to run crude machinery. However, cities were still not provided with adequate potable-water supplies or sanitary sewer facilities. The seventeenth-century Bills of Mortality of London posted shocking vital statistics that the death rate exceeded the birth totals in cities, while the reverse was true in rural England. Cities in the United States were no better off in the 1800s. Memphis had over 2000 deaths from yellow fever in 1873 and 5150 in 1878. Cholera was epidemic in both Paris and London, continuously ranging from bane to boon. It took these catastrophic events in major cities to stir the population and their elected officials to demand the installation of potable water and sewer systems.

Hamburg, Germany installed the first full-scale sewer system designed for collecting domestic sewage in 1843. In 1847 London, England issued a law requiring the draining of all privies into sewers. Memphis designed and constructed its first sanitary sewer system in 1880. Likewise, major American and European cities proceeded to construct sanitary sewers for the collection and transportation of domestic wastewater.

While many major cities had large storm sewers to drain clean water from built-up areas to avoid floods, draining sanitary wastewater into these pipes was prohibited. Paris was a prime example of this practice.

Baltimore banned the use of their 80,000 cesspools within the city in 1879 and fitted them with overflow piping for connection into the municipal storm sewer network. Brooklyn built sanitary sewers from 1857 to 1859, and Omaha had a complete sewer system designed by Colonel Waring in 1872.

The historic Hering Report to Boston in 1881 was based on Hering's survey of European sewer system practices, which has been called the "birth of modern sewerage works practices in the United States." Metcalf and Eddy, authors of the classic book on sewerage practices, published their first edition in 1914. They have recorded vital historical information and the state of the art for that period of time.

2-3 Empirical Methods

Early attempts at measurements were begun more than 3000 years ago by the Egyptian Pharaoh Menes, who developed a system of applied hydrology to help control the Nile's annual floods. The Nilometer, a simple staff gauge, the most basic of instruments, was used.

In premodern times, engineers based their designs for drainage systems on observations of storm events. Measurements of water volume were based on water flow from known areas in times of rain through natural channels, gutters, or watercourses with which they were familiar. Subsequently, the tributary areas could be precisely measured and were introduced as constants, and runoff estimates were based on a given depth of a respective amount of precipitation over the total district. However, on further research it was concluded that there is a gradual reduction in the immediate runoff per acre with an increase in the extent of the area or district; thus formulas were developed to account for this fact empirically. As time advanced, other factors became apparent, such as precipitation intensity, slope, soil conditions, and vegetation cover. The result in recent times has been a gradual development of more refined mathematical formulas.

According to the Metcalf and Eddy 1914 handbook entitled *American Sewerage Practice,* several empirical formulas were developed by leading engineers to estimate storm-water flows. Some of these classic equations are listed below for the reader's understanding of how present-day calculations are indebted to the previous work of these dedicated engineers.

Where Q = maximum discharge in cubic feet per second

i = maximum rate of rainfall in inches per hour (which is almost the same quantity as cubic feet per second per acre)

A = drainage area in acres

S = average slope of ground surface, in feet per 1000

C = constant

Hawksley (London), 1857: $Q = ACi \sqrt[4]{(S/Ai)}$

in which $C = 0.7$ and $i = 1.0$, so that $Q = 3.946\, A \sqrt[4]{S/A}$ (since $S = S/1000$).

Bürkli-Ziegler (Zurich), 1880: $Q = ACi \sqrt[4]{(S/A)}$

in which $C = 0.7$ to 0.9 and $i = 1$ to 3.

Adams (Brooklyn), 1880: $Q = ACi^{0.12}\sqrt{S/A^2 i^2}$

in which $C = 1.837$ and $i = 1$.

McMath (St. Louis), 1887: $Q = ACi^5\sqrt{S/A}$

in which $C = 0.75$ and $i = 2.75$.

Hering (New York), 1889:

$$Q = CiA^{0.85}S^{0.27} \quad \text{or} \quad Q\,Aci\,\sqrt[6]{S^{1.62}/A} = CiA^{0.833}S^{0.27}$$

in which Ci varies from 1.02 to 1.64.

Parmley (Cleveland), 1898: $\qquad Q = ACi\,\sqrt[6]{S^{1.5}/A}$

in which C is between 0 and 1 and $i = 4$ (intensity of rainfall for a period of 8 to 10 minutes; Walworth Sewer, Cleveland, used 4 in order to provide for most violent storm conditions).

Gregory (New York), 1907: $\qquad Q = ACiS^{0.186}/A^{0.14}$

in which $Ci = 2.8$ for impervious surfaces.

Allen Hazen prepared tables for the rapid application of the McMath formula which were published in the *American Civil Engineer's Pocket Book* (2d ed.) around the turn of the century.

Emil Kuichling tabulated various records and prepared curves of runoff showing rates of maximum flood discharges on certain American and English rivers under comparable conditions in the Mohawk Valley. This data was published in a report to the New York State Barge Canal in 1901. Kuichling's first formula for a curve gives rates of discharge which may be exceeded occasionally as

$$Q = \frac{44{,}000}{M + 170} + 20$$

The second formula for a curve gives rates of discharge which may be exceeded rarely as

$$Q = \frac{127{,}000}{M + 370} + 7.4$$

Kuichling's formula for floods which may be expected to occur frequently is

$$Q = \frac{25{,}000}{M + 125} + 15$$

where M = drainage area in square miles. Two other formulas were developed during that same period:

Murphy: $$Q = \frac{46,790}{M + 320} + 15$$

Metcalf and Eddy: $$Q = \frac{440}{M^{0.27}}$$

Other engineers developed subsequent formulas, not only for estimating floods but also for pipe. Both pressure and nonpressure conduits presented other problems for estimating.

Water contained in a closed conduit presented its own set of parameters, and the early engineers struggled with them for years.

Water seeks its own level. This level or surface is approximately perpendicular to the direction of the force of gravity. Conversely, if the water surface is not level, the water will flow from the higher elevation to the lower elevation. It's just another way of saying that the difference in pressure (or level) is what is necessary to make water flow and is technically known as "head."

If there is a certain difference in level (head) from upper to lower in a pipe, conduit, gutter, or channel, flow will be induced at a rate conditioned on fall as compared to the traversed length, cross-sectional area, surface roughness of conduit, whether pressure or nonpressure, whether flowing full or partially full, uniformity of flow, character, specific gravity, and viscosity of the liquid.

Many engineers and mathematicians developed formulas to address the understood parameters of estimating flow of water in various conduits. Bernoulli's landmark theorem was the basis for subsequent formulas. Hughes and Safford addressed critical velocity.

Men such as Darcy, Chezy, Bazin, Newell, Kutter, Jackson, Horton, Babb, Coffin, Cushman, Patch, Sherman, Schultz, Manning, Williams, Alword, Earl, Fuller, Hatton, Potter, Stevens, Hazen, and Ganguillet, to name just a few, contributed greatly to the refinement of formulas in use today.

In 1911 J. C. Stevens entered into a manufacturing agreement to make and market a new water level recorder of his own design. By 1914, he joined the firm Leupold & Stevens, where he became a well-known figure in the field of hydrology and hydraulic engineering. Many of his original recorders functioned for about 50 years.

Over time the basic hydraulic elements were understood and graphed. Data was tabulated and made available to the practicing engineer.

Weirs, flumes, float systems, and other measuring equipment with continuous recording have now become commonplace.

2-4   Modern Methods

Refinement of previous formulas and extended analyses over recent decades have made calculating water volumes and flows more accurate. The modern-day computer has speeded up the processing time.

Weather bureau records and their extensive gauging stations have become invaluable tools for determining rainfall intensities and estimating the 100-year flood elevation for specific areas contemplating bridge design, flood control, or other construction.

Earth satellites have provided accurate data for topography and aerial photography with space-age plotters produce superior mapping to assist in a hydraulic analysis of a given area.

Pipe material has been improved to reduce friction loss, and far more accurate factors for roughness have been introduced.

Flow measurement equipment and telemetry are now on the market as a result of space-age technology and the EPA-mandated requirements for wastewater treatment facility upgrading.

2-5   Present Usage of Water

The present usage of water in the industrialized sections of the world is at an all-time high and appears to be growing daily. Daily per capita averages vary from about 35 gallons in rural areas to over 200 gallons in some metropolitan settings, with the accepted mean amount of 100 gallons (per capita per day) generally used.

The term "sanitary sewage" is no longer adequate to all-inclusively describe the liquid now produced by urban living and discharged into municipal sewers. A new term was coined in the 1960s to define the modern city's dirty water: "wastewater."

The changes in character of modern wastewater have increased the need for more refined treatment facilities for cleanup so that it can be returned safely to the environment. Food wastes, laundry detergents, and other industrial compounds increase the burden of reclaiming the water. However, we must be aware that this all-encompassing mix of waste products into the community wastewater stream in all likelihood will increase.

As the population grows and industry expands, its thirst for process water and the need for pure drinking water and treatment become more acute. Additionally, as development spreads with the construction of more facilities, we need to consider avoiding building in the natural drainage patterns and wetland areas.

The last two or three decades have experienced an increasing public awareness of environmental concerns. Public water supply and wastewater treatment were greatly upgraded in the 1970s and 1980s through the funds provided by the United States Environmental Protection Agency (U.S. EPA).

2-6 Future Concerns

It is mandatory, in an age of shrinking individual space, that we solve the pressing problems of water resource management.

Protection of groundwater, preservation of wetlands, protection of rivers, adequate treatment of municipal wastewater, and control of discharge of industrial by-products into waterways are a must.

We have the resources and management capabilities to complete the work so seriously begun in the late 1970s. However, the financial burden may be our weakest link.

In summary, if prudent management of our water resources

does not keep pace with the increased demand, we certainly are in jeopardy of duplicating another "Dark Age."

2-7 Field Hydraulics

The foregoing portion of this manual has been devoted to laying a very brief overview for the reader. We are now at a point to begin our real purpose: the presentation of field hydraulics.

The definition of *field hydraulics,* as it pertains to this manual, is that data necessary for quick, reasonably accurate, and practical answers to a wide variety of hydraulic problems encountered in the field when access to reference material or sophisticated computers are not readily available.

The following sections contain very limited text, but provide tabulated data, basic formulas, charts, graphs, and illustrations for "on the spot" field solutions.

References

M. N. Cohn, *Sewers for Growing America,* Certainteed Products Corporation, 1966.

Metcalf and Eddy, *American Sewerage Practice,* McGraw-Hill, 1914.

R. O. Parmley, *Standard Handbook of Fastening and Joining,* McGraw-Hill, 1977

J. C. Stevens, *Water Resources Data Book,* 3d ed., Leupold & Stevens, April 1978.

World Book Encyclopedia, 1974 ed., "Hydraulics" and related articles.

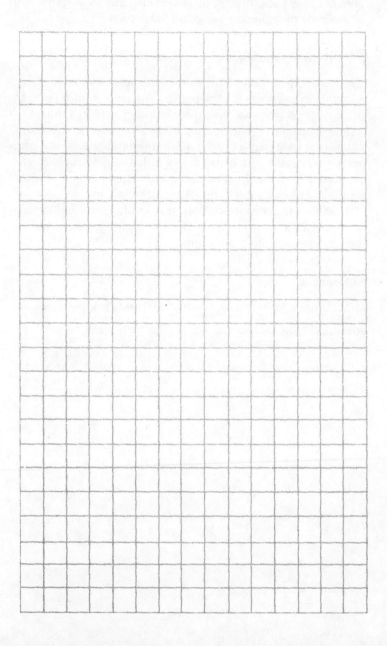

NOTES

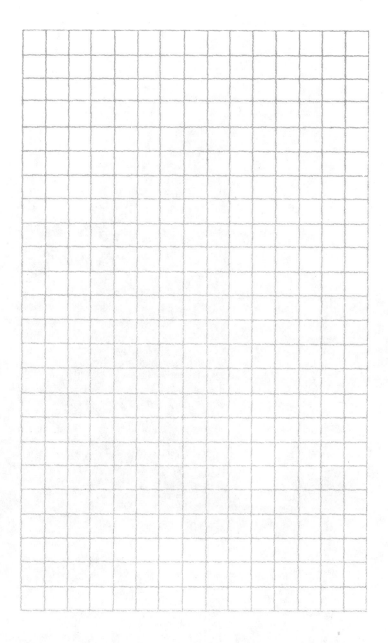

Groundwater

3-1   Origin and Occurrence

Groundwater is one segment of the earth's water circulation network or hydrologic cycle, previously discussed in detail, and has its origin in antiquity.

Groundwater is commonly understood to mean water occupying all the voids within a geologic formation. This saturated zone (water-filled area) is not to be confused with unsaturated, or aeration, zones where voids are filled with a mix of air and water or air only. Simply put, groundwater is the water below the surface of the earth that feeds wells and springs and helps to maintain lake levels in dry weather.

The groundwater (saturated) zone can be described as a gigantic natural reservoir or network of chambers within the earth's loose and discontinuous overburden of decayed rock debris whose capacity is the total volume of its openings or pores that are filled with water. Location of the groundwater zone in a specific site is governed by each respective geology, formation, movement, recharge, and other characteristics. For these reasons, it is impossible to adequately tabulate or even summarize all types of geologic environments where water may exist. Therefore, since this is not the purpose of this manual, it is suggested that the reader consult the professional literature for detailed specifics.

3-2   Modern Usage

Development of groundwater dates from ancient cultures and the Bible's Old and New Testaments make innumerable references to groundwater wells and springs. In fact, some of the construction methods of old still are in use, especially in the third world countries and areas of the Middle East.

In the continental United States, approximately 25 percent of the nation's domestic, industrial, and irrigation water originates from groundwater sources. About one-half of the United States' citizens get their potable-water supply from groundwater. Within the last three or four decades, reliance on groundwater has increased significantly.

Great technological advances have been made in groundwater technology since the turn of the century, but more significantly in recent years. Because of the limited scope of this manual, we are precluded from further discussion in this avenue. However, we advise that the reader consult some of the excellent handbooks and professional journals for enlightenment.

Daily consumption of water varies with age, area of the country, type of occupancy, and other factors. Refer to Table 3-1 for a general listing of daily water usage per capita or occupancy.

TABLE 3-1 Estimating Guidelines for Daily Water Usage

Category	Estimated water usage per day
Barber shop	100 gal per chair
Beauty shop	125 gal per chair
Boarding school, elementary	75 gal per student
Boarding school, secondary	100 gal per student
Clubs, civic	3 gal per person
Clubs, country	25 gal per person
College, day students	25 gal per student
College, junior	100 gal per student
College, senior	100 gal per student
Dentist's office	750 gal per chair
Department store	40 gal per employee
Drugstore	500 gal per store
Drugstore with fountain	2000 gal per store
Elementary school	16 gal per student
Hospital	400 gal per patient
Industrial plant	30 gal per employee + process water
Junior and senior high school	25 gal per student
Laundry	2000–20,000 gal
Launderette	1000 gal per unit
Meat market	5 gal per 100 ft² of floor area
Motel or hotel	125 gal per room
Nursing home	150 gal per patient
Office building	25 gal per employee
Physician's office	200 gal per examining room
Prison	60 gal per inmate
Restaurant	20–120 gal per seat
Rooming house	100 gal per tenant
Service station	600–1500 gal per stall
Summer camp	60 gal per person
Theater	3 gal per seat

Note that these quantities are general in nature and are not necessarily compatible with local or state regulatory code requirements. Therefore, they are considered here as guidelines only.

The historically accepted daily water usage per capita has been 100 gallons. However, studies and research by this writer have proved that communities vary significantly. Populations with a younger average age and children use far more water per capita than do rural elderly communities. Variations studied by this writer range from a high of 111 gallons per capita per day to a low of 45 gallons per capita per day. Therefore, it seems reasonable to assume a top average of 125 gallons per capita per day for domestic water usage when in preliminary planning for a municipal water system. This average includes a system loss factor.

3-3   Fundamentals of Flow

The basic fundamentals of groundwater flow were developed over a hundred years ago as a result of the "trial and error" method. Subsequent technical publications reported the empirical relationships with exact answers to specific situations. However, recent developments in soil mechanics, coupled with more sophisticated soil exploration processes and broad-scope computer mathematical capabilities, have given engineers greater theoretical insight into groundwater flow within the earth's soil media. Since these recent analytical techniques are complex, this manual will present only general concepts and basic data geared to field needs.

The movement of groundwater is affected by too many factors to delineate in this brief section. However, the following basics are presented to give field personnel an overview.

Infiltration.   Water may infiltrate into the upper layer of soil and ultimately into the groundwater. Its rate of infiltration depends on soil characteristics such as moisture content, slope, size of soil pores, vegetation, climatic condition, quantity of water, head, and ability of soil media to accept water.

Percolation.   Some of the water which infiltrates into the soil media may eventually percolate into the groundwater. The quantity of water available for percolation is that volume remaining after loss to the soil through the evapotranspiration process. Noted authorities estimate that approximately 20 percent of the water which infiltrates into the soil ultimately percolates into the groundwater.

Soil texture.   Soil components can be categorized into different general groups based on fineness or coarseness of its particles. Several accepted methods of classifying the particles by their respective size are in use today. These methods have been developed by the U.S. Bureau of Reclamation, AAOSHO, ASTM, Federal Aviation Authority, International Society of Soil Science, and the USDA. Refer to Fig. 3-1.

It should be noted that an individual soil will not consist entirely of clay, silt, or sand, but will be a blend or mix. Refer to Fig. 3-2 for various soil textural classes. The percentages of clay, silt, and sand are determined by mechanical analysis. Soil texture may also be estimated in the field by its physical characteristics.

Soil structure.   Soil structure relates to the aggregate's shape of soil particulates. Aggregates can have granular, prismatic, blocky, crumbly, or platey structures.

Porosity.   Porosity is the segment of a specific soil volume which is filled with either liquids or gases. Porosity is also influenced by the soil structure and generally varies with depth. Percent of pore space may be computed by the formula

$$PS = 100\% - \left(\frac{BD \cdot 100}{pd}\right)$$

FIGURE 3-1 USDA classification of soil based on particle size of 1 millimeter.

where PS = pore space percentage
        BD = bulk density
        pd = particle density

Permeability.   Permeability or hydraulic conductivity of the soil is a measurement of the ease with which water travels through a particular soil. The travel speed, or velocity, of water movement through a soil lens which is saturated can be related to permeability using the formula

$$V = K \frac{h}{D}$$

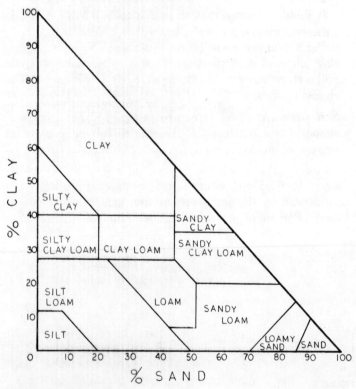

FIGURE 3-2 USDA soil textural classes.

where V = velocity of water flow in centimeters per second
k = saturated permeability in centimeters per second

$$\frac{h}{D} = \text{potential gradient}$$

Saturated permeability can be measured by many methods, either in the laboratory or in the field. Because of problems in duplicating or maintaining natural conditions in the lab, field measurements are generally more reliable.

Historically, the percolation test has been the standard test for determining suitability of site-specific soil for installation of absorption systems from sanitary septic tanks. However, results of percolation tests give a relative measure of soil permeability, but those results cannot be directly correlated with permeability. Some authorities, therefore, prefer the use of permeability. In the final analysis, the users of this manual should consult local and state regulatory codes for specific testing methods required for their respective areas.

In summary, groundwater is in a dynamic state and is moving from recharged areas to areas of discharge. The velocity of flow is very slow when compared to river current. As stated previously, groundwater movement is naturally controlled by two major factors: the permeability and the groundwater gradient.

3-4 Test Holes and Logs

Test holes or borings are made for a variety of reasons. However, they generally are made to obtain information. Therefore, it is extremely critical that their location and record of discovery be accurately logged. Refer to Fig. 3-3 for a typical example of a log.

Proper testing methods for specific test holes must always be obtained and understood to be in strict compliance with all applicable local, state, and federal codes and regulations. It is strongly recommended that anyone contemplating doing any drilling or boring into the groundwater be knowledgeable of all applicable codes and adequately skilled in the proper investigation techniques.

LOG OF WELL:

	Depths	Graphic Section	Rock Type	Color	Grain Size Mode	Grain Size Range	Miscellaneous Characteristics
	0-5		Silt	Dk yl bn	—	—	Siliceous. Much sand. Little gravel. Trace clay.
	5-10		Sand	"	C	Vfn/VC	Much gravel, silt. Trace clay.
D	10-15		"	"	S peb	Gran/M peb	Much silt. Trace gravel (including one VL pebble), clay.
R	15-20		Gravel	Mixed	S peb	Gran/M peb	Rhyolite,quartz sandstone,quartz,granite,trap. Mch sand,silt.
I	20-25		Sand	Yellow red	C/VC	Vfn/VC	Much silt. Little gravel.
F	25-30		"	"	C	"	Same.
T	30-35		Gravel	Mixed	S peb	Gran/VL peb	Rhyolite,quartz,granite,quartz sandstone,trap. Much sand,silt.
	35-40		Sand	Yellow red	C	Vfn/VC	Much gravel, silt.
	40-45		"	"	VC	"	Much gravel. Little silt.
	45-50		"	"	"	"	Much gravel, silt.
	50-55		Gravel	Mixed	S peb	Gran/L peb	Quartz, rhy, qtz ss, diorite, qrnt, trap. Mch sand, ltl silt.
	55-60		Sand	Yellow red	VC	Vfn/VC	Little gravel, silt.
	60-62		Gravel	Mixed	Gran	Gran/M peb	Granite, arkose, quartz, rhyolite, trap. Much sand, ltl silt.
	62-64		Sand	Yellow red	VC	Vfn/VC	Much gravel. Little silt.
	64-66		Gravel	Mixed	S peb	Gran/VL peb	Granite,rhyolite,quartz,granite schist,trap. Mch snd. Ltl silt
	66-68		Sand	Yellow red	C	Vfn/VC	Much gravel. Little silt.
	68-70		Gravel	Mixed	M peb	Gran/L peb	Rhyolite, granite, arkose, quartzite, trap. Mch sand, ltl silt.
72'	70-72		"	"	S peb	Gran/M peb	Same.

END OF LOG

FIGURE 3-3 Sample test hole log. (Courtesy: Morgan & Parmley, Ltd.)

32

3-5 Piezometric Mapping

A piezometric map illustrates the top elevation of the water table and can be used to determine the direction of groundwater flow.

Field information must be gathered to prepare a site-specific piezometric map. A series of borings must be made and depth to groundwater recorded at each location. Ground surface elevations at each respective bore hole must be accurately taken.

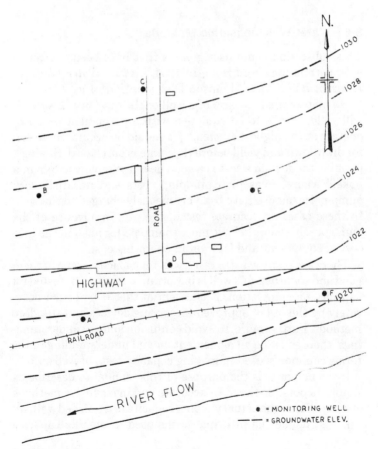

FIGURE 3-4 Generic piezometric map.

Precise horizontal location for each drill hole using the required coordinates must be field-measured. Finally, a general site reconnaissance must be made of the general area to plot natural and artificially constructed features and structures. Following the data collection, a map is prepared to show the groundwater contours relative to other site-specific features (see Fig. 3-4).

The water table is inclined in the direction of the flow and generally conforms roughly to the contour of the land surface. The gradient of the water table and the soil permeability control the water's velocity.

3-6   Water-Well Measuring Methods

This subsection contains segments that have been extracted, with permission, from Bulletin 1234, *Testing Water Wells for Drawdown and Yield,* Johnson Division, UOP, Inc.

No matter what the size of a well or the quantity of water it will yield, if it is to be equipped with a permanent pump to operate at the highest efficiency, it should be accurately tested for drawdown and yield before the pump is purchased. Buying a pump without such a test is a good deal like buying a "pig in a poke." Many times high pumping costs and unsatisfactory pump performance have been erroneously charged to the well. In these cases an accurate test of the well in advance of the pump purchase would have more than paid for itself in the first cost of equipment and later on in operating costs.

There are many ways to test a well for capacity, some of them good and some bad. If a well is important enough to be tested for capacity on completion, it is essential that it be tested accurately by the use of approved measuring devices and standard methods. Furthermore, to avoid confusion and misunderstandings, there must be an agreement on and understanding of the terms commonly used in making capacity tests of wells.

Inasmuch as it is the purpose of this manual to describe as simply as possible the most accurate and inexpensive methods of determining the actual yielding capacity of a finished well, we will first define the principal terms used in making capacity tests.

Definition of terms

If the meaning of each of the following terms is understood, any chance of confusion will be avoided.

1. *Static level.* This is the level to which water rises when the well is not being pumped. It is generally the level of the water table except in the case of artesian wells where the static level may be above the water table. In wells which do not flow, we can measure the static water level from the ground surface to the point where the water rises. Thus, when we say that the static level of a well is 50 feet, we mean that water stands 50 feet below the ground surface when there is no pumping. In wells which flow at the surface, the term "shut-in head" or simply "head" is commonly used to mean the head or pressure at the surface. It is not a term measuring quantity or volume but rather denotes pressure at the surface. If we say a well has a shut-in head or head of 10 feet at the surface, it means that water would rise 10 feet above the surface in a pipe.

2. *Pumping or dynamic level.* This is the level at which water stands in a well when pumping at any given rate. This level is variable and changes with the quantity being pumped. To determine this level, the same means are used as in measuring the static level. In important tests this level must be checked several times and no calculations of capacity should be made until several readings have been taken at least an hour apart. One can then be sure that the pumping level has become stationary or that the well has reached an equilibrium. As a matter of fact, the pumping level at any given rate of pumping will continue to lower for some time, but the amount of lowering is so small that it requires the use of extremely sensitive instruments to measure it. Even then the lowering may not be due to the pumping but to other causes over which no one has control. For all practical purposes, the pumping level can be determined when three readings of the level and capacity, taken one hour apart, are the same.

3. *Drawdown.* This is the distance the static level lowers under a given rate of pumping. It is the actual distance in feet between the static level and the pumping level.

Instruments used in testing wells

The gauges ordinarily used to measure pressure or head are the

1. Pressure-gauge reading in pounds per square inch.
2. Altitude-gauge reading in feet and fractions of a foot.
3. Vacuum-gauge reading in inches of vacuum or difference in pressure between mercury under atmospheric pressure and water being pumped.

Well drillers should familiarize themselves with these gauges and their uses. They are comparatively simple and easy to use, but for important tests should be calibrated with master gauges. It will facilitate reading and recording to have gauges whose range between zero and maximum reading will be approximately that of the quantities to be measured.

It is convenient to have handy the following table of conversions to facilitate the readings taken during a test and to ensure recording in the same units.

Unit	Pounds per square inch	Feet of water	Meters of water	Inches of mercury	Atmo-spheres
1 pound per square inch	1.0	2.31	0.704	2.04	0.0681
1 foot of water	0.433	1.0	0.305	0.882	0.02947
1 meter of water	1.421	3.28	1.00	2.89	0.0967
1 inch of mercury	0.491	1.134	0.3456	1.00	0.0334
1 atmosphere (sea level)	14.70	33.93	10.34	29.92	1.0000

Apparatus used in testing wells

For measuring a water level, the common unit is the foot, although it is sometimes necessary to take the readings in other units and convert them to feet for recording. For example, when a pressure gauge is used, the readings, which are in pounds pressure per square inch, must be multiplied by 2.31, the conversion factor for feet of water; and to convert a reading in feet to one in pounds per square inch, the reading is multiplied by 0.433. These are the two conversions most generally used.

Methods of measuring water levels

To make readings to any water level the simplest and also the crudest way is to use a heavy weight or float on the end of a steel tape or chalk line. Readings taken in this manner often vary from 2 to 6 inches from the true reading depending on the kind of tape used and whether the weighted end will float. The most satisfactory method is to use either an air line or an electric circuit. Measurements to the static or pumping level are always important, and the use of an air line or some other means more accurate than a weighted tape is generally required. Measurements to the pumping level are particularly important because they not only furnish the basis for estimating pumping levels for various capacities but also determine the power costs and the proper setting of the pump bowls if they are deep-well turbine type. Figure 3-5 shows a well in which an air line is installed to determine accurately the static level, pumping level, and drawdown. This apparatus is inexpensive, easy to install and remove, and generally accepted for all but the most accurate tests. It consists essentially of enough small-diameter pipe or tubing to extend from the surface well to a point below the static or pumping level depending on which is being measured. On top of the tubing or pipe is an ordinary altitude or pressure gauge to which is attached a tire pump or air pressure when it is available. Any small-size pipe of iron, brass, or copper, ⅛ or ¼ inch in diameter, may be used. If the air line is to be left in the well for permanent use, it should be of some noncorroding metal; if it is jointed and coupled, the joints must be absolutely airtight. Tubing is now commonly used for air lines. Two things are necessary to make accurate readings with this device.

1. The exact vertical distance between the center of the pressure or altitude gauge and the open end of the air line as it is installed in the well must be determined. This measurement is obtained by carefully measuring the air line as it is placed in the well. With pipe this can be done accurately; with tubing it

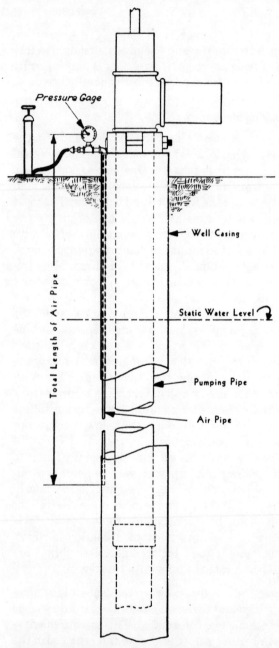

FIGURE 3-5 Air-line installation for measuring water levels. *(Courtesy: Johnson Division, UOP, Inc.)*

sometimes becomes a problem since the tubing has a tendency to curl. For this reason it is necessary to take special pains with tubing to calculate the vertical distance accurately.

2. The air line must be airtight from one end to the other, including the connections at the gauge and air pump.

Using air lines in testing wells

Be sure all joints and connections are airtight; then pump air into the line until the maximum possible pressure is reached. Air forced into the line creates pressure which forces the water out the lower end, leaving the line full of air. Readings should not be taken, of course, while the air is being pumped in. The head of water above the end of the line maintains the pressure and the gauge registers the actual pressure or head above the end of the line. A gauge graduated in feet shows directly the amount of submergence of the end of the line. Subtract this figure from the length of the line, and you have the water level. Just as drawdown is measured, so is recovery measured after pumping has stopped. In fact, recovery can be observed directly from the action of the needle as it moves to the right due to the column of water in the well rising to the static level when pumping is stopped. For active wells in good open aquifers this raise is very rapid, occurring within a few minutes. In poorer formations the recovery of static level may be a matter of hours. In any case, no well ever recovers the last few inches until a period of 24 to 72 hours has elapsed.

For a practical example, let us say that the gauge is an altitude gauge reading in feet and it takes 46 feet of head or pressure equivalent to this amount of head to force all of the water out of the air line. We have created a condition whereby the air pressure inside the air line is just sufficient to balance the pressure of the water on the outside of the line. In this case it is 46 feet by direct reading on the altitude gauge. If the distance from the center of the gauge to the open end of the air line is 95 feet, then the distance from the gauge to the static level is the difference between 95 feet and 46 feet, or 49 feet. If the gauge is above or below the ground surface, this must be considered if measurements are being referred to the surface.

Suppose now that we start our test pump. It will be noticed that the pressure on the gauge drops as the water is lowered in the well. This is as it should be since the water in the well is lowering and there is not as much water pressure above the open end of the air pipe as there was before pumping started. In most gauges the needle actually moves to the left, or counterclockwise, to indicate a drop in pressure. Most altitude gauges have two needles: one red and the other black. It is customary to set one needle, usually the red one, at the first reading — the reading taken before starting to pump the well. Then by reading the movable, or black, needle, one can read the difference between the two needles directly and determine the drawdown instantly. Let us say our moveable needle reads 22 feet after pumping has started. The difference between 46, our first reading, and 22 is 24 feet, or the drawdown at the quantity being pumped. A few trials will be sufficient for anyone to become familiar with the use of an altitude gauge and air line. If a pressure gauge is used, it operates exactly the same way except that the readings are in pounds per square inch and must be converted into feet of head for practical use.

Electrical methods for measuring

Perhaps the most popular device for measuring water levels is the electrical depth gauge or electric sounder. (See Fig. 3-6.) This tool is available from several manufacturers. A shielded electrode is suspended by a pair of insulated wires and a voltmeter indicates a flow of current when the electrode touches the water surface. Flashlight batteries supply the current. For accurate readings, the electrode and cable should be left hanging in the well between readings. This eliminates any error and kinks or bends in the wires which may change the length slightly when the device is pulled up and let down. For greater accuracy, the change in water level should be measured along the cable with a steel tape rather than using the metal markers which are usually attached to the cable by the manufacturer. One fixed mark may be used to indicate the static level and to serve as a base mark for subsequent measurements with the tape.

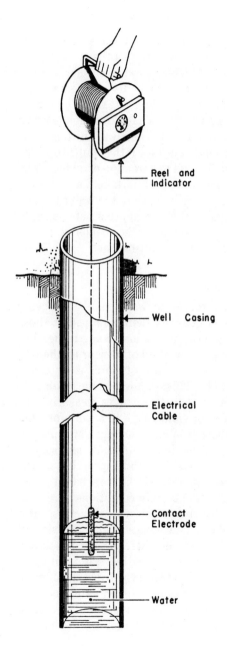

Reel and Indicator

Well   Casing

Electrical Cable

Contact Electrode

Water

FIGURE 3-6 Using electric sounder to measure water levels. *(Courtesy: Johnson Division, UOP, Inc.)*

Wetted-tape method

The wetted-tape method is a very accurate way of measuring depth of water and can be used readily for depths of up to 80 to 90 feet. First, a lead weight is attached to a steel measuring tape. The lower 2 or 3 feet of the tape is wiped dry and coated with carpenter's chalk or keel before a reading is taken. The tape is let down in the well until a part of the chalked section is below water, with one of the foot marks held exactly at the top of the casing or other measuring point. The tape is then pulled up. The wetted line on the tape can be read to a fraction of an inch, and this reading is subtracted from the footmark held at the measuring point to calculate the actual depth to the water level. A disadvantage of this method is that the approximate depth to water must be known so that a portion of the chalked section is submerged each time to produce a "wetting line."

Measuring quantity in testing wells

While the vertical measurements to water levels are being made, the measurements of quantity being pumped at those levels, which are just as important, are also being made. The most accurate and fundamental method of measuring the rate of flow of a steady continuous stream of water is by direct measure of volume and weight. This is known as the *volumetric method*. It must be accurately done. It involves measuring the volume of the pumped water by collecting it in a container and measuring the volume collected in a given length of time. For large quantities of water this method is unhandy and seldom used. Since this method is inconvenient and impractical for the average well job, other methods are used which are based on the behavior of flowing liquids under certain conditions of restriction. These methods involve the use of venturi meters, pitot tubes, current meters, orifices, flow gauges, weirs, and similar devices. The results are indirect and therefore not absolutely accurate, varying by small percentages either way from exact quantities. Most devices of this kind are easy to install but must be installed correctly and the readings taken carefully. The need for a compact, easily installed, and reasonably accurate measuring device of this type has led to the general acceptance and use of the circular-orifice weir. This device consists of a circular steel plate, $\frac{1}{16}$ inch thick, which is centered over the end of a discharge pipe, in which there is a perfectly circular hole with clean, square edges, smaller in diameter than the discharge pipe; and back 2 feet from this plate a small pipe ($\frac{1}{8}$ inch) is tapped smoothly at a right angle into the discharge pipe at the horizontal centerline.

The channel or pipe of approach should be at least 6 feet long overall. Two feet back from the end to which the orifice plate is attached, this discharge pipe should be tapped for $\frac{1}{8}$-inch pipe. All burrs, as a result of making this tap, should be carefully filed off on the inside of the pipe. A $\frac{1}{8}$-inch pipe nipple should be screwed into this hole and should be flush with the inside of the pipe.

The small tube in which the head of water is measured is called the *piezometer tube* and consists of a 5-foot length of rubber hose with a short section of glass tube fastened at one end and the opposite end attached to the $\frac{1}{8}$-inch pipe nipple that is screwed into the discharge pipe (see Fig. 3-7). The water level in this piezometer tube is kept visible in the glass by raising or lowering the end of the rubber hose.

The discharge pipe must be supported in a horizontal position using a level to make sure that it does not slant, and the piezometer connection must be in a straight line out from it when readings are taken. The pressure head on the orifice is measured as the vertical (up and down) distance from the level of the water in the piezometer tube down to the center of the orifice opening. The discharge pipe may be connected to the pump discharge by means of pipe connections, rubber or canvas hose, or other means which will conduct the water to the orifice and yet permit the orifice and channel of approach to be held in a rigid horizontal position. Water should be allowed to flow freely out of the piezometer tube until measurements are taken. This will eliminate from the tube line any obstruction such as sand, air bubbles, or other material.

The basic formula used in computation is Bernoulli's theorem: "At any section of a tube or pipe under steady flow without friction, the pressure head plus the velocity head is equal to the hydrostatic head that obtains when there is no

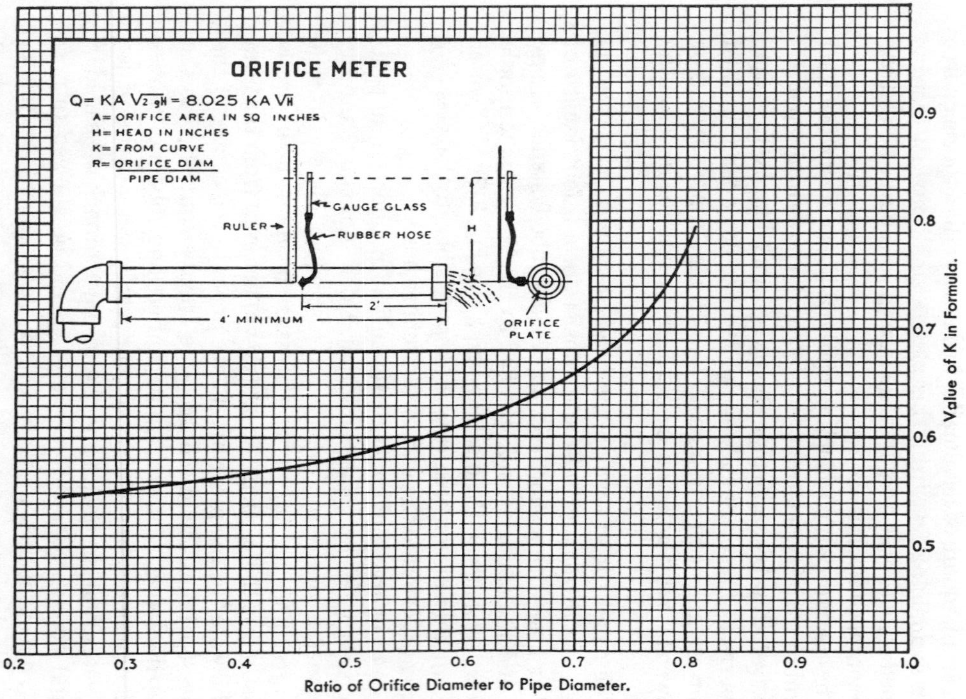
ORIFICE METER
Q= KA √2gH = 8.025 KA √H
A= ORIFICE AREA IN SQ INCHES
H= HEAD IN INCHES
K= FROM CURVE
R= ORIFICE DIAM / PIPE DIAM
GAUGE GLASS
RULER
RUBBER HOSE
H
2'
4' MINIMUM
ORIFICE PLATE
0.2
0.3
0.4
0.5
0.6
0.7
0.8
0.9
1.0
Ratio of Orifice Diameter to Pipe Diameter.
0.5
0.6
0.7
0.8
0.9
Value of K in Formula.

FIGURE 3-7 Curve for determining value of *K* in formula for circular-orifice weir. *(Courtesy: Johnson Division, UOP, Inc.)*

flow." The contraction of the stream flowing through an orifice of this kind naturally affects the amount of discharge. It is customary, therefore, to correct the theoretical discharge by a number or constant called the *coefficient of discharge*. In the basic formula of $Q = KA \sqrt{2gH}$, the error is less than three-tenths of 1 percent when the head on the center of the orifice is greater than two times its diameter (Merriman). For most cases, then, the actual discharge from a circular vertical orifice of area A may be computed from $Q = KA\sqrt{2gH} = K \times$ $8.02 \times A\sqrt{h}$, where the acceleration due to gravity is 32.2 feet per second, $2g = 64.4$ and $\sqrt{2g} = 8.02$, K is the coefficient of discharge and H the head on the orifice, Q is in gallons per minute (gpm), and A is in square inches. The values of K are taken from the curve in Figure 3-7 describing the device.

This method has certain limitations which must be considered.

1. The pipe on which the orifice is used must be horizontal and the discharge fall free.

2. The edges of the orifice opening must be sharp and clean, preferably chamfered to 45° with the sharp edge upstream.

3. Combinations of pipe and orifice diameters must be such that the head built up will be at least three times the diameter of the orifice.

4. The orifice must be vertical and centered in the discharge pipe.

5. The head measuring tube must be free from air bubbles and not protrude beyond the inside surface of the pipe.

The apparatus and setup are shown in Fig. 3-7 for all ordinary tests. For absolute accuracy each individual orifice should be calibrated before use.

Estimating yields greater than tests

Generally, after drawdown and capacity have been determined, the question arises as to the quantity that may be obtained at different drawdowns up to the maximum or 100 percent of the water in the well. The common method of making such estimates has been to assume that drawdown and yield are directly proportional and if a well pumped 100 gpm at 10-foot drawdown, it would pump 200 gpm at 20-foot drawdown. A much safer way to estimate the yield of wells beyond their tested capacity is to apply the rule governing the relationship between drawdown and yield. This means considering first whether the well is a nonartesian or artesian well and checking the type of formation from which it obtains water. Figure 3-8 shows these relationships for either type of well constructed in unconsolidated formations. A little study with a few trials in the field will soon convince the most skeptical person of both the practicability of this method of predicting yields and drawdown and its essentially conservative nature. The mistake most generally made in connection with testing wells is erroneously estimating

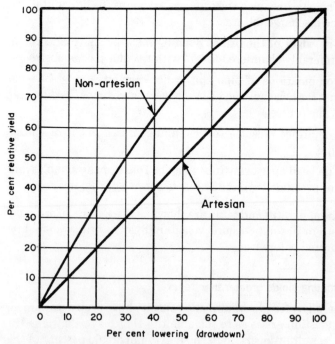

FIGURE 3-8 Relation of drawdown to yield in water table and artesian wells. *(Courtesy: Johnson Division, UOP, Inc.)*

capacity at drawdowns greater than those obtained during the testings. It should be understood that only certain wells yield in direct proportion to drawdown. For properly designed artesian wells the relationship is that of direct proportion, but for non-artesian wells or those constructed in free, unconfined water strata the relationship is a curve with the relationship varying rapidly after drawdowns of 50 percent of the water depth have been reached.

To grasp the use of the curve, consider that the maximum capacity of a well, or a yield of 100 percent, is reached when there is no more water in the well, in other words, when the lowering of water is equal to 100 percent of the original depth of water in the well. Let us take a well constructed in an unconsolidated formation in which there is 100 feet of water in a free or nonartesian state. If we lower the water by pumping 25 feet, it is the same as a lowering of 25 percent, and so on. Suppose we are actually testing a well and obtaining 100 gpm with a lowering of 20 feet. We find from the curve that with this amount of lowering we are developing about 36 percent of the maximum capacity or yield at 100 percent drawdown. So, if 100 gpm is 36 percent, then the maximum or 100 percent is 100/0.36, or about 278 gallons. Suppose we want to know how much we would be getting with a drawdown of 30 feet or lowering of 30 percent of the original depth of water. By consulting the curve, we find that 30 percent lowering develops about 52 percent of the maximum yield, or 52 percent of 278, which is 145 gallons. Any combination of drawdown or yield can be figured the same way. This simple method can also be applied to pump setting problems, and the proper pump setting for any desired capacity within the depth of water available can be calculated with reasonable accuracy once a careful capacity test has been made. It should be remembered that it is considered bad practice to pump water below the top of the bowls of a turbine pump or the top of a well screen.

The proper testing of a well is no easy job. It requires time and equipment to be done properly. The information obtained from accurate tests is worth many times its cost; therefore, a well owner should not hesitate to require a new well to be tested and should be willing to pay for it.

It should also be kept in mind that a pump test is a test of the finished well as a structure and not necessarily a test of the yielding ability of the formation in which the well is constructed. This is because the formation may have an ability to yield several times the amount of water pumped from the well. In actual practice many wells are so constructed that they cannot possibly make the full yield of a formation available. This is why the proper selection of slot opening, length of screen, diameter of screen, and method of well construction used are all vitally important for any new well.

A discussion of the procedures for testing wells to obtain data for calculating the principal factors of aquifer performance is beyond the scope of this manual. In such tests, the variation of drawdown with time of pumping must be determined. Measurements of drawdown in one or more observation wells are required. After the pumping is stopped, the recovery of water levels in the pumped well and observation wells is measured.

3-7   Springs

A spring is a localized discharge of groundwater occurring at the ground surface in the form of flowing water. Springs are not to be confused with seepage areas which emit or weep water at a much slower discharge than a spring.

Springs exist in many forms and have been classified as to cause, discharge, temperature, variability, and rock structure. Some springs are the result of gravitational forces, while others are a product of nongravitational energy such as produced by volcanic action.

Groundwater flowing to the ground surface producing springs is a result of hydrostatic pressure. The following types have been classified:

1. *Depression springs.*   Formed at intersections of ground surface and water table.

2. *Contact springs.*   Developed by permeable water-bearing aquifer overlying a less permeable aquifer which intersects at the surface.

3. *Artesian springs.*   Point of release of water which is under

pressure from a confined aquifer either at an outcrop of the aquifer or via an opening in the confining bed.

4. *Impervious rock springs.* Erupting from fractures or conduit channels within impervious rock formations.

5. *Tubular or fracture springs.* Bursting forth from circular channels, such as solution channels or lava tubes in impermeable rock which is linked to groundwater.

Most springs fluctuate in their rate of discharge, which is in response to their rate of recharge.

3-8 Geothermal Conditions

Geothermal conditions within the earth produce thermal springs which discharge water having temperatures in excess of local normal groundwater. This hydrothermal phenomenon releases water and steam as a by-product of volcanic activity.

Geysers are thermal springs that periodically spout steam and hot water resulting from the expansive force of superheated steam within the earth's subsurface channels. Water from surface sources drains down into deep vertical openings where it is heated well above the boiling point which expands, producing steam which accelerates the need to find an escape, finally finding energy release into the atmosphere at ground level.

Field investigation of these conditions should be executed under the most careful means possible. Highly qualified technical personnel should be in charge of all field surveys where these geothermal conditions exist. Therefore, no attempt is made in this manual to provide the necessary methods.

3-9 Artificial Recharge

Artificial recharge of the groundwater is the result of man's involvement in the natural order of things.

According to reliable authorities, recharging of groundwater began in Europe early in the nineteenth-century and in the United States near the end of that century.

A variety of methods have been employed to artificially recharge the groundwater. The following paragraphs briefly summarize the most commonly practiced methods.

1. *Stream channel.*   This method involves water spreading in a natural stream channel which increases the time and overall area where water is recharging the groundwater from a naturally bottom lossing channel. This, of course, requires upstream management of stream flow and channel alterations to heighten the desired infiltration.

2. *Basin.*   Water placed into an artificially built basin or reservoir, generally constructed with earthen dikes or berms, and allowed to seep from the bottom into the groundwater. Multiple basins are provided for operation continuity.

3. *Ridge and furrow.*   Water is distributed to a series of furrows (ditches) that are shallow, flat-bottomed, and evenly spaced to obtain maximum water-contact area. Three styles of construction are utilized. One follows the contour of the existing ground by means of sharp switchbacks which meander to and fro across the land. Another design is tree-shaped with a main canal which successively branches into smaller canals and ditches. The remaining type is lateral where a series of small furrows extend laterally from a primary canal.

4. *Flooding.*   Perhaps the most cost-effective method of artificial recharge is the flooding of flatland which generally needs little or no construction, just a means to deliver water to the site. Best results of infiltration occur on sites with undisturbed vegetation and soil covering with water delivery slow to avoid erosion.

5. *Irrigation.*   Irrigation provides water to croplands during the growing season. Excess water during winter and dormant periods can be delivered cheaply because the distribution facilities are already in existence.

6. *Pit.*   Pits are excavated in permeable formations, but are costly. Gravel pits that have been abandoned are ideal for this method.

7. *Recharge well.*   A recharge well is described as a well that injects water from the surface into the groundwater aquifer. This process is the opposite of pumping a well, but construction may or may not be similar. As water is pumped into the aquifer,

a cone of recharge is formed which is similar in shape to, but a reverse of, the cone of depression which surrounds a normal pumping well.

8. *Storm sewers.* Storm sewers and snow-melt that is drained or partially diverted by artificially constructed grader ditches, storm sewers, and curb and gutter facilities contribute to the collection of waters; which, in turn, concentrate this runoff in areas that periodically recharge the groundwater.

9. *Wastewater reuse.* Reuse of wastewater is increasing in many parts of the world as the population expands and fresh water becomes more difficult to obtain.

In the late 1970s and early 1980s, the U.S. EPA encouraged municipalities to design and construct wastewater treatment facilities with effluent land disposal via seepage cells. This alternative method of discharge protected the nation's surface waters while artificially recharging the groundwater.

In order to promote this technology, EPA grants were increased if a community constructed this type of treatment facility. See Figure 3-9 for a basic cross-sectional view of a generic effluent seepage cell. Note that a mound will usually be formed in the water table beneath a land disposal facility. The height and lateral extent of the rise in the saturated zone will vary

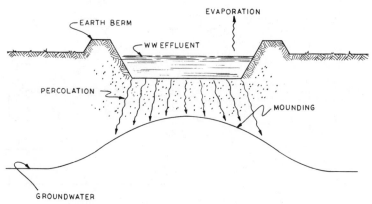

FIGURE 3-9 **Geometry of a wastewater effluent seepage cell.**

according to many conditions which are too numerous to discuss here.

However, a basic formula for estimating the height of a groundwater mound which results from an artificial recharge is

$$H = \left(\frac{Q \log (R/r)}{1.3 \, C} + h^2 \right)^{1/2}$$

where Q = flow in gallons per day
C = coefficient of permeability in gallons per day per square foot
H = see Fig. 3-10
h = see Fig. 3-10
R = see Fig. 3-10
r = see Fig. 3-10

3-10   Unconfined Aquifer

An unconfined aquifer is one condition where the water table varies in form, as well as slope, depending on the areas of recharge-discharge, well pumpage, and permeability. Rise and fall

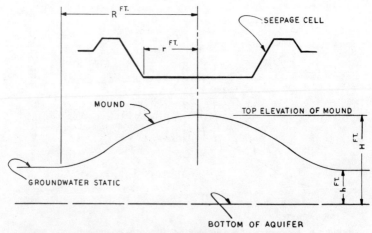

FIGURE 3-10   Basic geometry of a typical mounding effect.

of the groundwater table is directly related to changes in the volume of water stored within a specific aquifer. Refer to Fig. 3-11 for a cross-sectional view illustrating unconfined and confined aquifers.

A special class of an unconfined aquifer is the perched water condition. This situation occurs wherever a groundwater segment is isolated from the main body of groundwater by an impermeable stratum or lens of relatively small area. Wells drilled into these limited water sources yield only temporary or small quantities of water and, therefore, must be avoided for permanent water supplies or limited in their pumpage. Refer to Fig. 3-12 for an illustrative sketch.

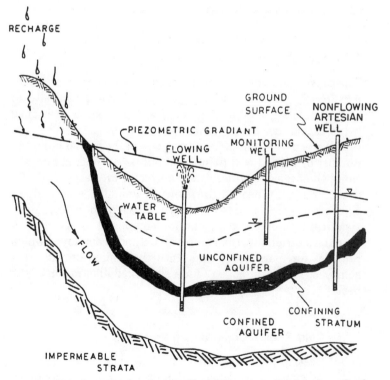

FIGURE 3-11 Cross-sectional view illustrating unconfined and confined aquifers.

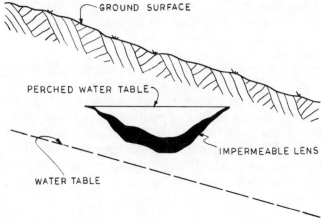

FIGURE 3-12 Perched water table.

3-11 Construction Dewatering

The construction dewatering to be considered here is a tempo-rary lowering of the existing groundwater, at a specific site, to enable installation of a component which must be positioned below the surface of the normal water table.

The size of the dewatering pump, size of the header pipe, and number of needed well points will not be considered here be-cause the project engineer and contractor are responsible for determining the needed drawdown. Generally, a specialized contractor is employed to execute this phase of the work, since a significant capital investment for equipment is usually re-quired. Additionally, the empirical method is employed which provides for more well points until the desired drawdown is achieved. Refer to Fig. 3-13 for a simplified illustration of the basics.

3-12 Cold-Region Construction Problems

Snow, sleet, ice, and frost are conditions that exist in cold re-gions and cause many construction problems. Engineers must take all of these factors and their effects into consideration when designing and constructing a project in cold regions.

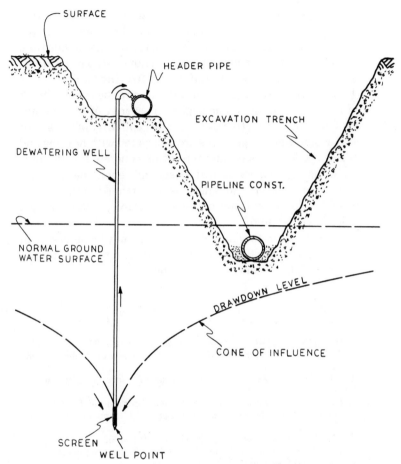

FIGURE 3-13 **Typical construction dewatering arrangement.**

While it is not the purpose of this manual to address all of these conditions, it is paramount to call attention to moisture in the upper soil strata which can result in frozen ground during cold seasons.

Frozen ground basically is very strong and will provide a good structural support for short periods, but long-term load creeping will result, especially in warm ice-rich conditions.

The most critical frost problems result from the freeze-thawing cycles. If footings are not below the frost line, they have a distinct possibility to shift. Road and street pavements experience cracking, frost boils, and general deterioration if the subbase material does not allow adequate drainage of moisture. Sidewalks and similar components will not be stable, if water is present in the subbase material during freezing weather.

Heaving is the result of water migration to the freezing front and ice-lens formation. The degree of severity of heaving has been found to be directly related to the capillarity of the soil or subbase material. Large forces are associated with heaving and can frost-jack large piles and reposition concrete foundations.

In arctic regions, permafrost exists and general design principles recommend that foundations be extended to the permanently frozen ground and provide a suitable, stable thermal barrier in the ground.

Extensive laboratory and in situ testing may be necessary in addition to a conventional survey of soil conditions in order to develop designs to control frost action.

References

J. S. Ameen, *Source Book of Community Water Systems,* 1960.

F. G. Driscoll, *Groundwater and Wells,* 2d ed., Johnson Filtration Systems, 1986.

E. Eranti and G. C. Lee, *Cold Region Structural Engineering,* McGraw-Hill, 1986.

Guideline Document: Design, Construction, and Operation of Land Disposal Systems for Liquid Wastes, Wisconsin Dept. of Natural Resources, Nov. 1975.

M. E. Harr, *Groundwater and Seepage,* McGraw-Hill, 1962.

Morgan & Parmley, Ltd., Professional Consulting Engineers, project records, miscellaneous files, and plans.

D. K. Todd, *Groundwater Hydrology,* 2d ed., Wiley, 1980.

NOTES

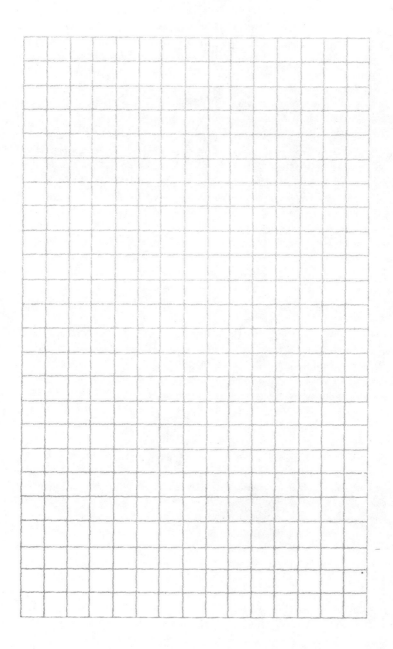

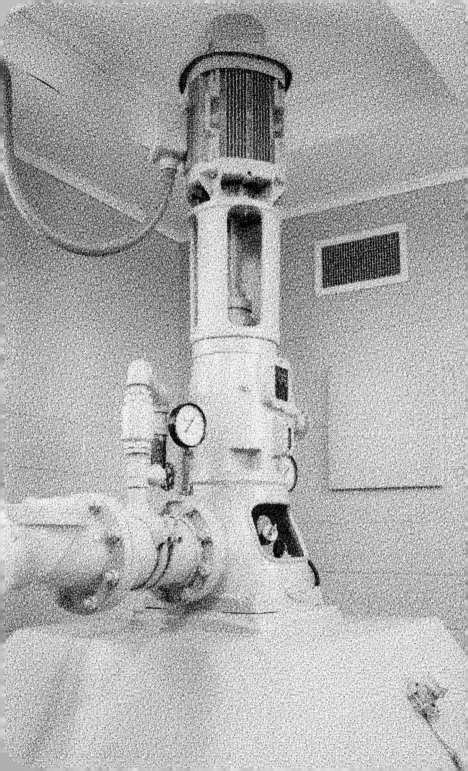

Pumps

TABLE 4-1 Major Pump Types and Construction Styles

Pump Type And Construction Style	Distinguishing Construction Characteristics	Used Orientation	Used No. Of Stages	Relative Maintenance Requirement	Comments
Dynamic					
Centrifugal					
Horizontal					
Single stage overhung, process type	Impeller cantilevered beyond bearing.	Horizontal	1	Low	Most common style used in process service.
Two stage overhung, process type	2 impellers cantilevered beyond bearing.	"	2	Low	For heads above single stage capability.
Single stage impeller between bearing	Impeller between bearings; casing radially or axially split.	"	1	Low	For high flows to 330 m head
Chemical	Casing patterns designed with thin sections for high cost alloys; small sizes.	"	1	Medium	Low pressure and temperature ratings.
Slurry	Large flow passage, erosion control features.	"	1	High	Low speed; adjustable axial clearance.
Canned	Pump and motor enclosed in pressure shell; no stuffing box.	"	1	Low	Low head-capacity limits for models used in chemical services.
Multistaged, horizontally split casing	Nozzles, usually in bottom half of casing.	"	Multi	Low	For moderate temperature-pressure ratings.
Multistage, barrel type	Outer casing confines inner stack of diaphragm.	"	Multi	Low	For high temperature-pressure ratings.
Vertical					
Single stage process type	Vertical orientation.	Vertical	1	Low	Style used primarily to exploit low NPSH requirement.
Multistage, process type	Many stages, low head/stage.	"	Multi	Medium	High head capability, low NPSH requirement.
In-line	Arranged for in-line installation, like a valve	"	-1	Low	Allows low cost installation, simplified piping systems.
High Speed	Speeds to 380 rps, head to 1770 m	"	1	Medium	Attractive cost for high head/flow flow.
Sump	Casing immersed in sump for installation convenience and priming ease.	"	1	Low	Low cost installation.
Multistage deep well	Very long shafts	Vertical	Multi	Medium	Water well service with driver at grade.
Axial (Propeller)	Propeller shaped impeller, usually larger size.	"	1	Low	A few applications in chemical plants and refineries.
Turbine (Regenerative)	Fluted impeller; flow path like screw around periphery.	Horizontal	1,2	Med. to High	Low flow-high head performance. Capacity virtually independent of head.
Positive Displacement					
Reciprocating					
Piston, plunger	Slow speeds; valves, cylinders, stuffing boxes subject to wear.	Horizontal	1	High	Driven by steam engine cylinders or motors through crankcase.
Metering	Small units with precision flow control system	"	1	Medium	Diaphragm and packed plunger types.
Diaphragm	No stuffing box; can be pneumatically or hydraulically actuated.	"	1	High	Used for chemical slurries; diaphragms prone to failure.
Rotary					
Screw	1, 2 or 3 screw rotors	"	1	Medium	For high viscosity, high flow high pressure.
Gear	Intermeshing gear wheels	"	1	Medium	For high viscosity, moderate pressure, moderate flow.

SOURCE: *Fluid Flow Pocket Handbook* by Nicholas P. Cheremisinoff. Copyright 1984, Gulf Publishing Co., Houston, Texas. Used with permission. All rights reserved.

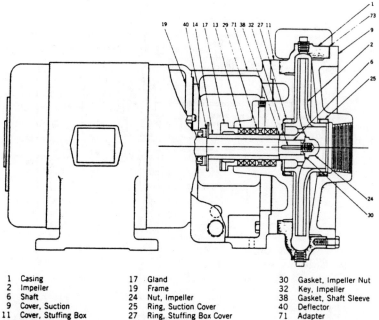

1	Casing	17	Gland	30	Gasket, Impeller Nut
2	Impeller	19	Frame	32	Key, Impeller
6	Shaft	24	Nut, Impeller	38	Gasket, Shaft Sleeve
9	Cover, Suction	25	Ring, Suction Cover	40	Deflector
11	Cover, Stuffing Box	27	Ring, Stuffing Box Cover	71	Adapter
13	Packing	29	Ring, Lantern	73	Gasket
14	Sleeve, Shaft				

The numbers shown on this drawing *do not* necessarily represent standard part numbers in use by any manufacturer.

FIGURE 4-1 Overhung impeller, close-coupled, single-stage, end-suction pump. *(Courtesy: Hydraulics Institute.)*

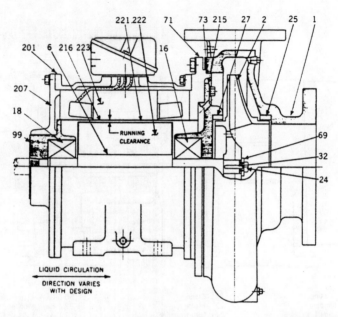

1	Casing	27	Ring, Stuffing Box	207	Cover, Motor End
2	Impeller	32	Key, Impeller	215	Housing Bearing and
6	Shaft, Pump	69	Lockwasher		Wearing Ring
16	Bearing, Inboard	71	Adapter	216	Can, Stator
18	Bearing, Outboard	73	Gasket	221	Can, Rotor
24	Nut, Impeller	99	Housing, Bearing	222	Assembly, Rotor Core
25	Ring, Suction Cover	201	Housing, Stator	223	Assembly, Stator Core

The numbers shown on this drawing do not necessarily represent standard part numbers in use by any manufacturer. The cross sectional drawings illustrate the largest possible number of parts in their proper relationship and a few construction modifications but do not necessarily represent recommended design.

FIGURE 4-2 Overhung impeller, close-coupled, single-stage, end-suction, canned motor pump. *(Courtesy: Hydraulics Institute.)*

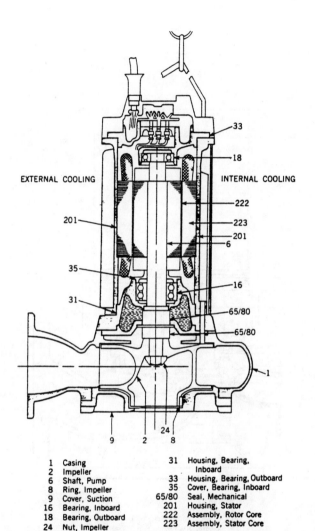

EXTERNAL COOLING   INTERNAL COOLING

1	Casing
2	Impeller
6	Shaft, Pump
8	Ring, Impeller
9	Cover, Suction
16	Bearing, Inboard
18	Bearing, Outboard
24	Nut, Impeller
31	Housing, Bearing, Inboard
33	Housing, Bearing, Outboard
35	Cover, Bearing, Inboard
65/80	Seal, Mechanical
201	Housing, Stator
222	Assembly, Rotor Core
223	Assembly, Stator Core

The numbers shown on this drawing *do not* necessarily represent standard part numbers in use by any manufacturer. The cross sectional drawings illustrate the largest possible number of parts in their proper relationship and a few construction modifications but *do not* necessarily represent recommended design.

FIGURE 4-3 Overhung impeller, close-coupled, single-stage, submersible pump. *(Courtesy: Hydraulics Institute.)*

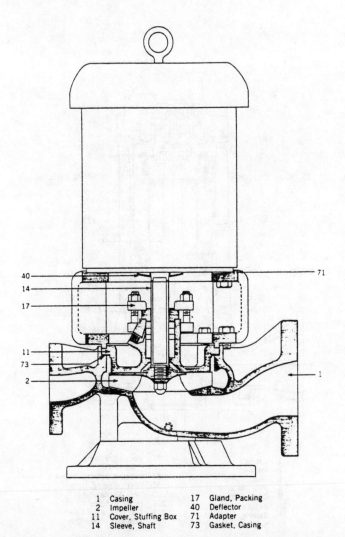

1	Casing	17	Gland, Packing
2	Impeller	40	Deflector
11	Cover, Stuffing Box	71	Adapter
14	Sleeve, Shaft	73	Gasket, Casing

The numbers shown on this drawing *do not* necessarily represent standard part numbers in use by any manufacturer. The cross sectional drawings illustrate the largest possible number of parts in their proper relationship and a few construction modifications but *do not* necessarily represent recommended design.

FIGURE 4-4 Overhung impeller, close-coupled, single-stage, inline pump. *(Courtesy: Hydraulics Institute.)*

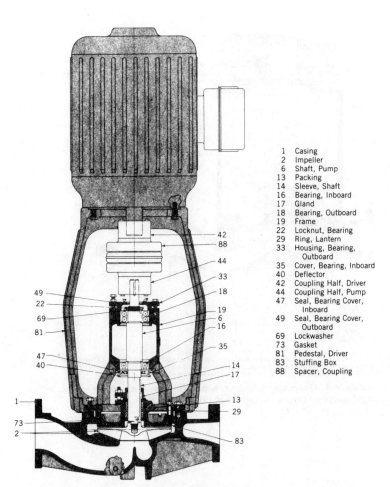

1	Casing
2	Impeller
6	Shaft, Pump
13	Packing
14	Sleeve, Shaft
16	Bearing, Inboard
17	Gland
18	Bearing, Outboard
19	Frame
22	Locknut, Bearing
29	Ring, Lantern
33	Housing, Bearing, Outboard
35	Cover, Bearing, Inboard
40	Deflector
42	Coupling Half, Driver
44	Coupling Half, Pump
47	Seal, Bearing Cover, Inboard
49	Seal, Bearing Cover, Outboard
69	Lockwasher
73	Gasket
81	Pedestal, Driver
83	Stuffing Box
88	Spacer, Coupling

The numbers shown on this drawing *do not* necessarily represent standard part numbers in use by any manufacturer. The cross sectional drawings illustrate the largest possible number of parts in their proper relationship and a few construction modifications but *do not* necessarily represent recommended design.

FIGURE 4-5 Overhung impeller, separately coupled, single-stage, inline, flexible coupling pump. *(Courtesy: Hydraulics Institute.)*

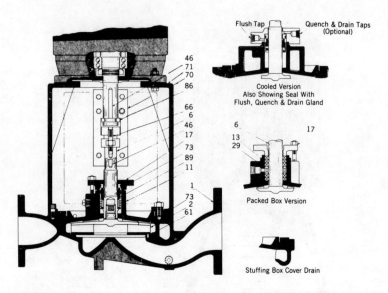

1	Casing	17	Gland	70	Coupling, Shaft
2	Impeller	29	Ring, Lantern	71	Adapter
6	Shaft, Pumping	46	Key, Coupling	73	Gasket
11	Cover, Stuffing Box	61	Plate, Side	86	Ring, Thrust, Split
13	Packing	66	Nut, Shaft Adjusting	89	Seal

The numbers shown on this drawing *do not* necessarily represent standard part numbers in use by any manufacturer. The cross sectional drawings illustrate the largest possible number of parts in their proper relationship and a few construction modifications but *do not* necessarily represent recommended design.

FIGURE 4-6 Overhung impeller, separately coupled, single-stage, inline, rigid coupling pump. *(Courtesy: Hydraulics Institute.)*

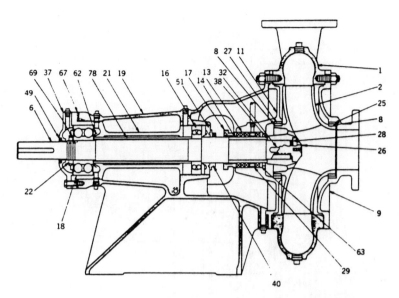

1	Casing	19	Frame	38	Gasket, Shaft Sleeve
2	Impeller	21	Liner, Frame	40	Deflector
6	Shaft, Pump	22	Locknut, Bearing	49	Seal, Bearing Cover,
8	Ring, Impeller	25	Ring, Suction Cover		Outboard
9	Cover, Suction	26	Screw, Impeller	51	Retainer, Grease
11	Cover, Stuffing Box	27	Ring, Stuffing Box Cover	62	Thrower (Oil or Grease)
13	Packing	28	Gasket, Impeller Screw	63	Bushing, Stuffing Box
14	Sleeve, Shaft	29	Ring, Lantern	67	Shim, Frame Liner
16	Bearing, Inboard	32	Key, Impeller	69	Lockwasher
17	Gland	37	Cover, Bearing,	78	Spacer, Bearing
18	Bearing, Outboard		Outboard		

The numbers shown on this drawing *do not* necessarily represent standard part numbers in use by any manufacturer.

FIGURE 4-7 Overhung impeller, separately coupled, single-stage, frame-mounted pump. *(Courtesy: Hydraulics Institute.)*

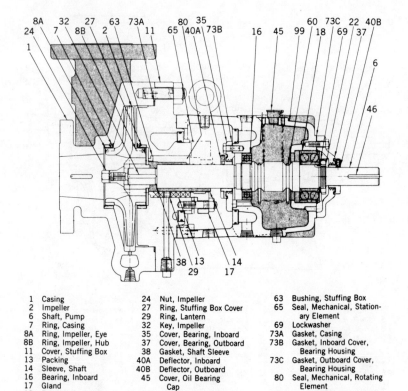

1	Casing	24	Nut, Impeller	63	Bushing, Stuffing Box
2	Impeller	27	Ring, Stuffing Box Cover	65	Seal, Mechanical, Station-
6	Shaft, Pump	29	Ring, Lantern		ary Element
7	Ring, Casing	32	Key, Impeller	69	Lockwasher
8A	Ring, Impeller, Eye	35	Cover, Bearing, Inboard	73A	Gasket, Casing
8B	Ring, Impeller, Hub	37	Cover, Bearing, Outboard	73B	Gasket, Inboard Cover,
11	Cover, Stuffing Box	38	Gasket, Shaft Sleeve		Bearing Housing
13	Packing	40A	Deflector, Inboard	73C	Gasket, Outboard Cover,
14	Sleeve, Shaft	40B	Deflector, Outboard		Bearing Housing
16	Bearing, Inboard	45	Cover, Oil Bearing	80	Seal, Mechanical, Rotating
17	Gland		Cap		Element
18	Bearing, Outboard	46	Key, Coupling	99	Housing, Bearing
22	Locknut, Bearing	60	Ring, Oil		

The numbers shown on this drawing *do not* necessarily represent standard part numbers in use by any manufacturer.

FIGURE 4-8 Overhung impeller, separately coupled, single-stage center-line support, API 610 pump. *(Courtesy: Hydraulics Institute.)*

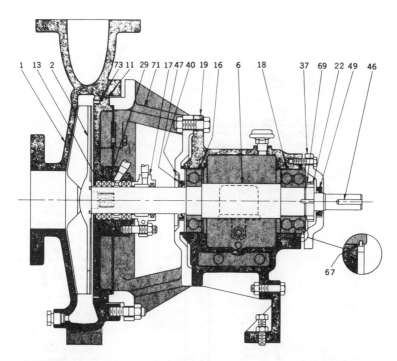

1	Casing	18	Bearing, Outboard	47	Seal, Bearing Cover,
2	Impeller	19	Frame		Inboard
6	Shaft, Pump	22	Locknut, Bearing	49	Seal, Bearing, Cover,
11	Cover, Stuffing Box	29	Ring, Lantern		Outboard
13	Packing	37	Cover, Bearing, Outboard	67	Shim, Frame Liner
16	Bearing, Inboard	40	Deflector	69	Lockwasher
17	Gland	46	Key, Coupling	71	Adapter
				73	Gasket

The numbers shown on this drawing *do not* necessarily represent standard part numbers in use by any manufacturer.

FIGURE 4-9 Overhung impeller, separately coupled, single-stage, frame-mounted, ANSI B73-1 pump. *(Courtesy: Hydraulics Institute.)*

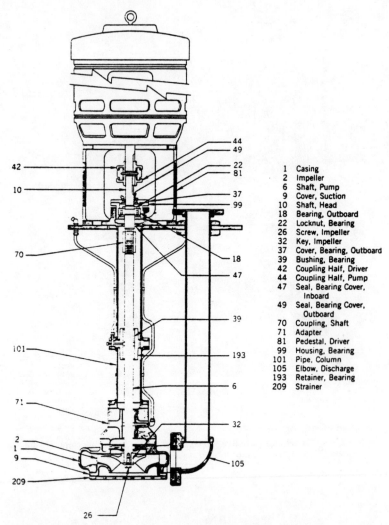

1	Casing
2	Impeller
6	Shaft, Pump
9	Cover, Suction
10	Shaft, Head
18	Bearing, Outboard
22	Locknut, Bearing
26	Screw, Impeller
32	Key, Impeller
37	Cover, Bearing, Outboard
39	Bushing, Bearing
42	Coupling Half, Driver
44	Coupling Half, Pump
47	Seal, Bearing Cover, Inboard
49	Seal, Bearing Cover, Outboard
70	Coupling, Shaft
71	Adapter
81	Pedestal, Driver
99	Housing, Bearing
101	Pipe, Column
105	Elbow, Discharge
193	Retainer, Bearing
209	Strainer

The numbers shown on this drawing *do not* necessarily represent standard part numbers in use by any manufacturer.

FIGURE 4-10   Overhung impeller, separately coupled, single-stage, wet pit volume pump. *(Courtesy: Hydraulics Institute.)*

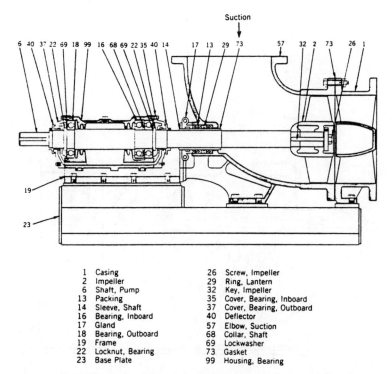

Suction

6 40 37 22 69 18 99 16 68 69 22 35 40 14   17 13 29 73   57   32 2 73   26 1

19

23

1	Casing	26	Screw, Impeller
2	Impeller	29	Ring, Lantern
6	Shaft, Pump	32	Key, Impeller
13	Packing	35	Cover, Bearing, Inboard
14	Sleeve, Shaft	37	Cover, Bearing, Outboard
16	Bearing, Inboard	40	Deflector
17	Gland	57	Elbow, Suction
18	Bearing, Outboard	68	Collar, Shaft
19	Frame	69	Lockwasher
22	Locknut, Bearing	73	Gasket
23	Base Plate	99	Housing, Bearing

The cross sectional drawings illustrate the largest possible number of parts in their proper relationship and a few construction modifications but *do not* necessarily represent recommended design.

FIGURE 4-11   Axial-flow horizontal pump. *(Courtesy: Hydraulics Institute.)*

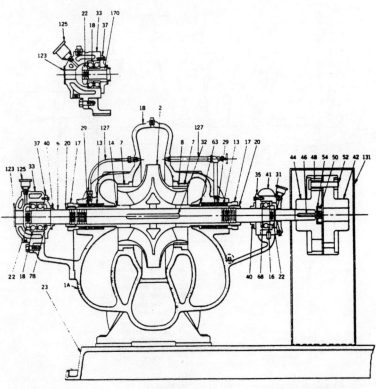

1A	Casing, Lower Half	23	Base Plate	46	Key, Coupling
1B	Casing, Upper Half	29	Ring, Lantern	48	Bushing, Coupling
2	Impeller	31	Housing, Bearing, Inboard	50	Locknut, Coupling
6	Shaft, Pump	32	Key, Impeller	52	Pin, Coupling
7	Ring, Casing	33	Housing, Bearing, Outboard	54	Washer, Coupling
8	Ring, Impeller			63	Bushing, Stuffing Box
13	Packing	35	Cover, Bearing, Inboard	68	Collar, Shaft
14	Sleeve, Shaft	37	Cover, Bearing, Outboard	78	Spacer, Bearing
16	Bearing, Inboard	40	Deflector	123	Cover, Bearing End
17	Gland	41	Cap, Bearing, Inboard	125	Cup, Grease
18	Bearing, Outboard	42	Coupling Half, Driver	127	Piping, Seal
20	Nut, Shaft Sleeve	44	Coupling Half, Pump	131	Guard, Coupling
22	Locknut			170	Adapter, Bearing

The numbers shown on this drawing *do not* necessarily represent standard part numbers in use by any manufacturer.

FIGURE 4-12   Impeller between bearings, separately coupled, single-stage axial (horizontal) split case pump (part one). *(Courtesy: Hydraulics Institute.)*

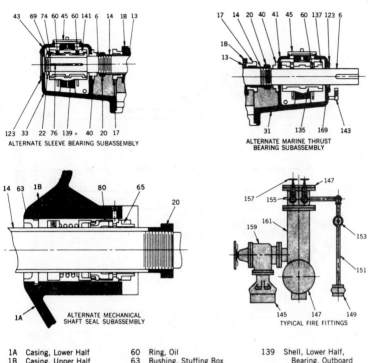

1A	Casing, Lower Half	60	Ring, Oil	139	Shell, Lower Half,
1B	Casing, Upper Half	63	Bushing, Stuffing Box		Bearing, Outboard
6	Shaft, Pump	65	Seal, Mechanical,	141	Shell, Upper Half,
13	Packing		Stationary Element		Bearing, Outboard
14	Sleeve, Shaft	69	Lockwasher	143	Gauge, Sight, Oil
17	Gland	74	Journal, Thrust Bearing	145	Cone, Discharge, Large
20	Nut, Shaft Sleeve	76	Key, Bearing Journal	147	Flange, Blank
22	Locknut, Bearing	80	Seal, Mechanical, Rotating	149	Cone, Discharge, Small
31	Housing, Bearing, Inboard		Element	151	Pipe, Test
33	Housing, Bearing,	123	Cover, Bearing End	153	Valve, Test
	Outboard	135	Shell, Lower Half,	155	Manifold, Hose Valve
40	Deflector		Bearing, Inboard	157	Valve, Hose
41	Cap, Bearing, Inboard	137	Shell, Upper Half,	159	Valve, Relief
43	Cap, Bearing, Outboard		Bearing, Inboard	161	Fitting, Discharge
45	Cover, Oil, Bearing Cap			169	Seal, Bearing Housing

The numbers shown on this drawing *do not* necessarily represent standard part numbers in use by any manufacturer.

FIGURE 4-13 Impeller between bearings, separately coupled, single-stage axial (horizontal) split case pump (part two). *(Courtesy: Hydraulics Institute.)*

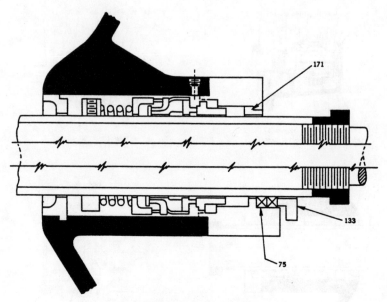

ALTERNATE MECHANICAL SHAFT SEAL SUBASSEMBLY
WITH OPTIONAL THROTTLE BUSHING
OR AUXILIARY STUFFING BOX

75	Stuffing Box, Auxiliary	171	Bushing, Throttle,
133	Gland, Stuffing Box, Auxiliary		Auxiliary

The numbers shown on these drawings *do not* necessarily represent standard part numbers in use by any manufacturer.

FIGURE 4-14   Impeller between bearings, separately coupled, single-stage axial (horizontal) split case pump (part three). *(Courtesy: Hydraulics Institute.)*

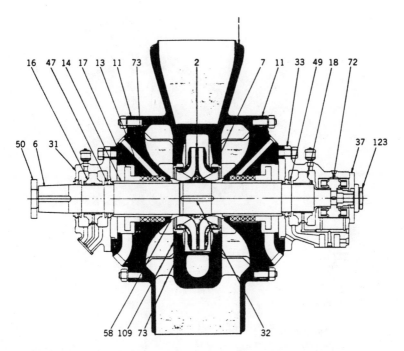

1	Casing	17	Gland	49	Seal, Bearing Cover,
2	Impeller	18	Bearing, Outboard		Outboard
6	Shaft, Pump	31	Housing, Bearing, Inboard	50	Locknut, Coupling
7	Ring, Casing	32	Key, Impeller	58	Sleeve
11	Cover, Stuffing Box	33	Housing, Bearing,	72	Collar, Thrust
13	Packing		Outboard	73	Gasket, O-Ring
14	Sleeve, Shaft	37	Cover, Bearing, Outboard	109	Diaphragm
16	Bearing, Inboard	47	Seal, Bearing Cover,	123	Cover, Bearing End
			Inboard		

The numbers shown on this drawing *do not* necessarily represent standard part numbers in use by any manufacturer.

FIGURE 4-15   Impeller between bearings, separately coupled, single-stage radial (vertical) split case pump. *(Courtesy: Hydraulics Institute.)*

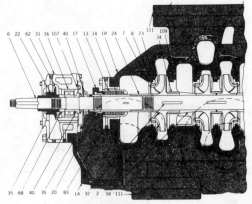

1A	Casing, Lower Half	22	Locknut, Bearing	63	Bushing, Stuffing Box
1B	Casing, Upper Half	24	Nut, Impeller	68	Collar, Shaft
2	Impeller	31	Housing, Bearing, Inboard	72	Collar, Thrust
5	Diffuser	32	Key, Impeller	73	Gasket
6	Shaft, Pump	33	Housing, Bearing, Outboard	83	Stuffing Box
7	Ring, Casing	34	Sleeve, Impeller Hub	107	Shield, Oil Retaining
8	Ring, Impeller	35	Cover, Bearing, Inboard	109	Diaphragm, Interstage
13	Packing	37	Cover, Bearing, Outboard	111	Crossover, Interstage
14	Sleeve, Shaft	40	Deflector	113	Bushing, Interstage Diaphragm
16	Bearing, Inboard	45	Cover, Oil, Bearing Cap	115	Ring, Balancing
17	Gland	56	Disc or Drum, Balancing	117	Bushing, Pressure Reducing
18	Bearing, Outboard	58	Sleeve, Interstage	119	Coupling, Oil Pump
20	Nut, Shaft Sleeve	62	Thrower (Oil or Grease)	121	Pump, Oil

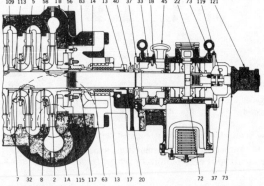

The numbers shown on this drawing *do not* necessarily represent standard part numbers in use by any manufacturer.

FIGURE 4-16   Impeller between bearings, separately coupled, multistage axial (horizontal) split case pump. *(Courtesy: Hydraulics Institute.)*

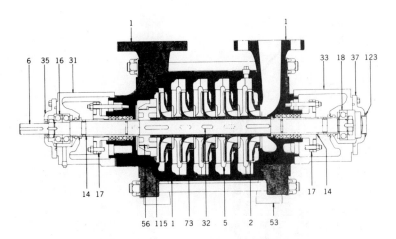

1	Casing	33	Housing, Bearing, Outboard
2	Impeller		
5	Diffuser	35	Cover, Bearing, Inboard
6	Shaft, Pump	37	Cover, Bearing, Outboard
14	Sleeve, Shaft	53	Base
16	Bearing, Inboard	56	Drum, Balancing
17	Gland	73	Gasket, O-Ring
18	Bearing, Outboard	115	Ring, Balancing
31	Housing, Bearing, Inboard	123	Cover, Bearing End
32	Key, Impeller		

The numbers shown on this drawing *do not* necessarily represent standard part numbers in use by any manufacturer.

FIGURE 4-17 Impeller between bearings, separately coupled, multistage radial (vertical) split case pump. *(Courtesy: Hydraulics Institute.)*

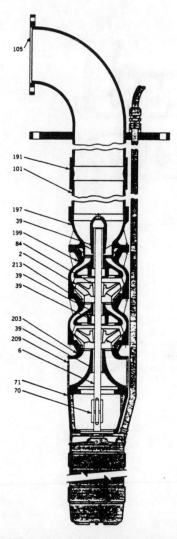

2	Impeller
6	Shaft, Pump
39	Bushing, Bearing
70	Coupling, Shaft
71	Adapter
84	Collet, Impeller Lock
101	Pipe, Column
105	Elbow, Discharge
191	Coupling, Column Pipe
197	Case, Discharge
199	Bowl, Intermediate
203	Case, Suction
209	Strainer
213	Ring, Bowl

The cross sectional drawings illustrate the largest possible number of parts in their proper relationship and a few construction modifications but *do not* necessarily represent recommended design.

FIGURE 4-18   Turbine-type, vertical, multistage, deep-well, submersible pump. *(Courtesy: Hydraulics Institute.)*

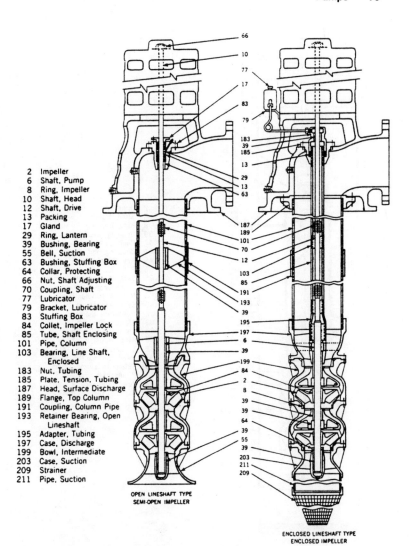

2	Impeller
6	Shaft, Pump
8	Ring, Impeller
10	Shaft, Head
12	Shaft, Drive
13	Packing
17	Gland
29	Ring, Lantern
39	Bushing, Bearing
55	Bell, Suction
63	Bushing, Stuffing Box
64	Collar, Protecting
66	Nut, Shaft Adjusting
70	Coupling, Shaft
77	Lubricator
79	Bracket, Lubricator
83	Stuffing Box
84	Collet, Impeller Lock
85	Tube, Shaft Enclosing
101	Pipe, Column
103	Bearing, Line Shaft, Enclosed
183	Nut, Tubing
185	Plate, Tension, Tubing
187	Head, Surface Discharge
189	Flange, Top Column
191	Coupling, Column Pipe
193	Retainer Bearing, Open Lineshaft
195	Adapter, Tubing
197	Case, Discharge
199	Bowl, Intermediate
203	Case, Suction
209	Strainer
211	Pipe, Suction

OPEN LINESHAFT TYPE
SEMI-OPEN IMPELLER

ENCLOSED LINESHAFT TYPE
ENCLOSED IMPELLER

The cross sectional drawings illustrate the largest possible number of parts in their proper relationship and a few construction modifications but do not necessarily represent recommended design.

FIGURE 4-19 Turbine-type, vertical, multistage, deep-well pump. *(Courtesy: Hydraulics Institute.)*

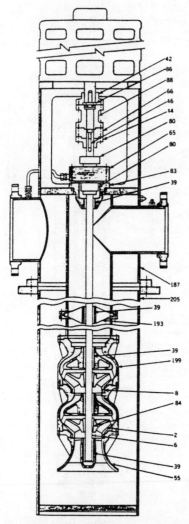

2	Impeller
6	Shaft, Pump
8	Ring, Impeller
39	Bushing, Bearing
42	Coupling Half, Driver
44	Coupling Half, Pump
46	Key, Coupling
55	Bell, Suction
65	Seal, Mechanical, Stationary Element
66	Nut, Shaft Adjusting
80	Seal, Mechanical, Rotating Element
83	Stuffing Box
84	Collet, Impeller Lock
86	Ring, Thrust, Split
88	Spacer, Coupling
187	Head, Surface Discharge
193	Retainer, Bearing, Open Lineshaft
199	Bowl, Intermediate
205	Barrel or Can Suction

The cross sectional drawings illustrate the largest possible number of parts in their proper relationship and a few construction modifications but *do not* necessarily represent recommended design.

FIGURE 4-20 Turbine-type, vertical, multistage, barrel or can pump. *(Courtesy: Hydraulics Institute.)*

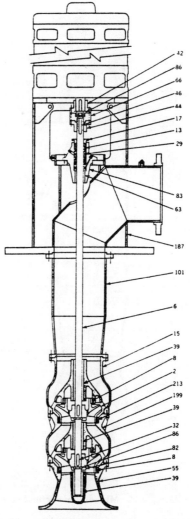

2	Impeller
6	Shaft, Pump
8	Ring, Impeller
13	Packing
15	Bowl, Discharge
17	Gland
29	Ring, Lantern
32	Key, Impeller
39	Bushing, Bearing
42	Coupling Half, Driver
44	Coupling Half, Pump
46	Key, Coupling
55	Bell, Suction
63	Bushing, Stuffing Box
66	Nut, Shaft Adjusting
82	Ring, Thrust, Retainer
83	Stuffing Box
86	Ring, Thrust, Split
101	Pipe, Column
187	Head, Surface Discharge
199	Bowl, Intermediate
213	Ring, Bowl

The cross sectional drawings illustrate the largest possible number of parts in their proper relationship and a few construction modifications but *do not* necessarily represent recommended design.

FIGURE 4-21 Turbine-type, vertical, multistage, short setting pump. *(Courtesy: Hydraulics Institute.)*

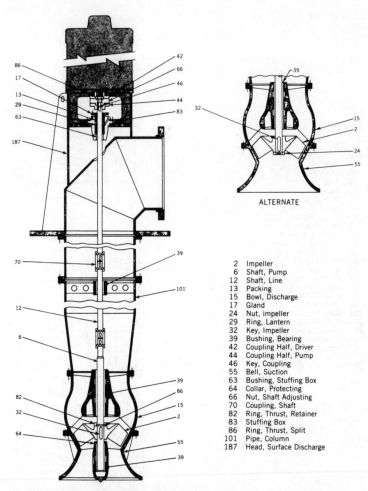

2   Impeller
6   Shaft, Pump
12   Shaft, Line
13   Packing
15   Bowl, Discharge
17   Gland
24   Nut, impeller
29   Ring, Lantern
32   Key, Impeller
39   Bushing, Bearing
42   Coupling Half, Driver
44   Coupling Half, Pump
46   Key, Coupling
55   Bell, Suction
63   Bushing, Stuffing Box
64   Collar, Protecting
66   Nut, Shaft Adjusting
70   Coupling, Shaft
82   Ring, Thrust, Retainer
83   Stuffing Box
86   Ring, Thrust, Split
101   Pipe, Column
187   Head, Surface Discharge

The cross sectional drawings illustrate the largest possible number of parts in their proper relationship and a few construction modifications but *do not* necessarily represent recommended design.

FIGURE 4-22   **Mixed-flow vertical pump.** *(Courtesy: Hydraulics Institute.)*

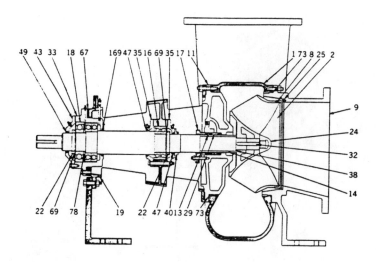

1	Casing	22	Locknut, Bearing	47	Seal, Bearing Cover,
2	Impeller	24	Nut, Impeller		Inboard
8	Ring, Impeller	25	Ring, Suction Cover	49	Seal, Bearing Cover,
9	Cover, Suction	29	Ring, Lantern		Outboard
11	Cover, Stuffing Box	32	Key, Impeller	67	Shim, Frame Liner
13	Packing	33	Housing, Bearing,	69	Lockwasher
14	Sleeve, Shaft		Outboard	73	Gasket
16	Bearing, Inboard	35	Cover, Bearing, Inboard	78	Spacer, Bearing
17	Gland	38	Gasket, Shaft Sleeve	169	Seal, Bearing Housing
18	Bearing, Outboard	40	Deflector		
19	Frame	43	Cap, Bearing Outboard		

The numbers shown on this drawing *do not* necessarily represent standard part numbers in use by any manufacturer.

FIGURE 4-23 **Overhung impeller, separately coupled, single-stage, mixed-flow impeller volute-type horizontal pump.** *(Courtesy: Hydraulics Institute.)*

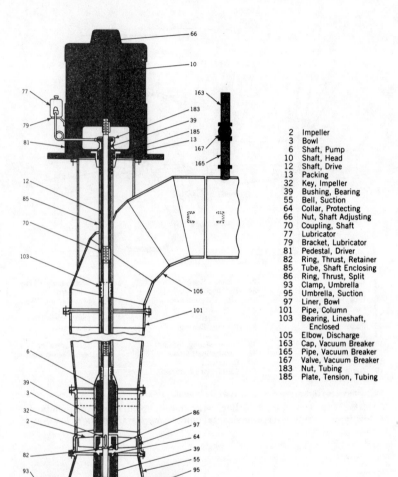

2   Impeller
3   Bowl
6   Shaft, Pump
10 Shaft, Head
12 Shaft, Drive
13 Packing
32 Key, Impeller
39 Bushing, Bearing
55 Bell, Suction
64 Collar, Protecting
66 Nut, Shaft Adjusting
70 Coupling, Shaft
77 Lubricator
79 Bracket, Lubricator
81 Pedestal, Driver
82 Ring, Thrust, Retainer
85 Tube, Shaft Enclosing
86 Ring, Thrust, Split
93 Clamp, Umbrella
95 Umbrella, Suction
97 Liner, Bowl
101 Pipe, Column
103 Bearing, Lineshaft,
      Enclosed
105 Elbow, Discharge
163 Cap, Vacuum Breaker
165 Pipe, Vacuum Breaker
167 Valve, Vacuum Breaker
183 Nut, Tubing
185 Plate, Tension, Tubing

The cross sectional drawings illustrate the largest possible number of parts in their proper relationship and a few construction modifications but *do not* necessarily represent recommended design.

FIGURE 4-24   Vertical, axial-flow impeller (propeller)–type pump. *(Courtesy: Hydraulics Institute.)*

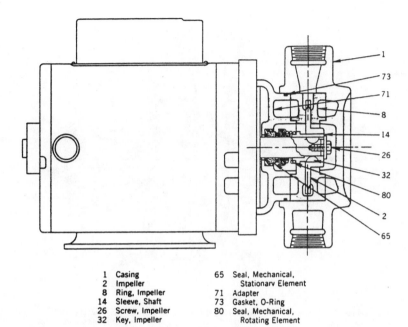

1	Casing	65	Seal, Mechanical,
2	Impeller		Stationary Element
8	Ring, Impeller	71	Adapter
14	Sleeve, Shaft	73	Gasket, O-Ring
26	Screw, Impeller	80	Seal, Mechanical,
32	Key, Impeller		Rotating Element

The cross sectional drawings illustrate the largest possible number of parts in their proper relationship and a few construction modifications but *do not* necessarily represent recommended design.

FIGURE 4-25 Regenerative turbine, impeller overhung, single-stage pump. *(Courtesy: Hydraulics Institute.)*

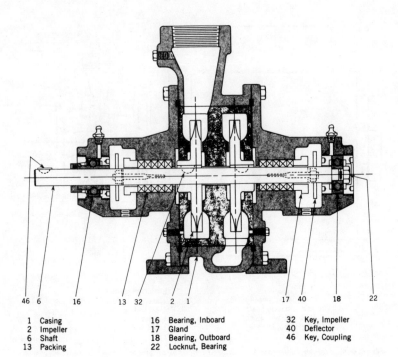

1	Casing	16	Bearing, Inboard	32	Key, Impeller
2	Impeller	17	Gland	40	Deflector
6	Shaft	18	Bearing, Outboard	46	Key, Coupling
13	Packing	22	Locknut, Bearing		

The cross sectional drawings illustrate the largest possible number of parts in their proper relationship and a few construction modifications but *do not* necessarily represent recommended design.

FIGURE 4-26 Regenerative turbine, impeller between bearings, two-stage pump. *(Courtesy: Hydraulics Institute.)*

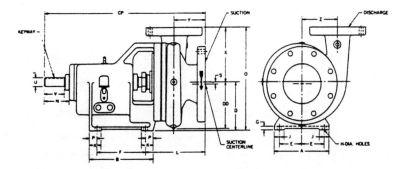

A — Width of base support.
B — Length of base support.
CP — Length of pump.
D — Vertical height—bottom of base support to centerline of pump.
DD — Distance—pump centerline to bottom drain plug.
E — Distance from centerline pump to centerline hold-down bolts.
F — Distance from centerline to centerline of hold-down bolt holes.
G — Thickness of pads on support, or height of base plate, depending on location of bolt holes.
H — Diameter of hold-down bolt holes.
J — Width of pads for hold-down bolts.
K — Length of support pad for hold-down bolts.
L — Horizontal distance from suction nozzle face to centerline nearest hold-down bolt holes.

N — Distance—end of bearing housing to end of shaft.
O — Vertical distance—bottom of support to discharge nozzle face or top of case on horizontally split pumps.
P — Length from edge of support, or base plate, to centerline of bolt holes.
S — Distance from centerline of pump to centerline of suction nozzle.
U — Diameter of straight shaft—coupling end.
V — Length of shaft available for coupling or pulley.
X — Distance from discharge face to centerline of pump.
Y — Horizontal distance—centerline discharge nozzle to suction nozzle face.
Z — Centerline discharge nozzle to centerline of pump.

Note: Where multiple dimensions for similar components are required, i.e., mounting pad widths and locations, subscripts 1, 2, 3, et cetera should be used. Number from right to left, i.e., HE_1, HE_2, HE_3. These subscript designations may appear in a view other than indicated.

FIGURE 4-27 Overhung impeller, separately coupled, single-stage, frame-mounted pump. *(Courtesy: Hydraulics Institute.)*

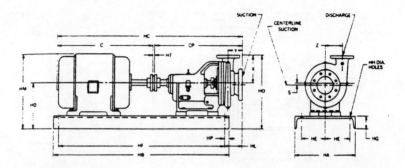

C — Length of driver.
CP — Length of pump.
HA — Width of base support.
HB — Length of base support.
HC — Overall length of combined pump and driver when on base.
HD — Vertical height—bottom of base support to centerline of pump.
HE — Distance from centerline pump to centerline hold-down bolts.
HF — Distance from centerline to centerline of hold-down bolt holes.
HG — Thickness of pads on support, or height of base plate, depending on location of bolt holes.
HH — Diameter of hold-down bolt holes.
HL — Horizonal distance from suction nozzle face to centerline nearest hold-down bolt holes.

HM — Height of unit—bottom of base to top of driver.
HO — Vertical distance—bottom of support to discharge nozzle face or top of case on horizontally split pumps.
HP — Length from edge of support, or base plate, to centerline of bolt holes.
HT — Horizontal distance—between pump and driving shaft.
S — Distance from centerline of pump to centerline of suction nozzle.
X — Distance from discharge face to centerline of pump.
Y — Horizontal distance—centerline discharge nozzle to suction nozzle face.
Z — Centerline discharge nozzle to centerline of pump.

Note: Where multiple dimensions for similar components are required, i.e., mounting pad widths and locations, subscripts 1, 2, 3, et cetera should be used. Number from right to left, i.e., HE$_1$, HE$_2$, HE$_3$. These subscript designations may appear in a view other than indicated.

FIGURE 4-28 Overhung impeller, separately coupled, single-stage, frame-mounted pump on baseplate. *(Courtesy: Hydraulics Institute.)*

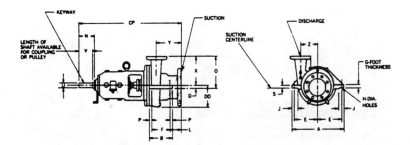

A — Width of base support.
B — Length of base support.
CP — Length of pump.
D — Vertical height—bottom of base support to centerline of pump.
DD — Distance—pump centerline to bottom drain plug.
E — Distance from centerline pump to centerline hold-down bolts.
F — Distance from centerline to centerline of hold-down bolt holes.
G — Thickness of pads on support, or height of base plate, depending on location of bolt holes.
H — Diameter of hold-down bolt holes.
J — Width of pads for hold-down bolts.
L — Horizontal distance from suction nozzle face to centerline nearest hold-down bolt holes.

N — Distance—end of bearing housing to end of shaft.
O — Vertical distance—bottom of support to discharge nozzle face or top of case on horizontally split pumps.
P — Length from edge of support, or base plate, to centerline of bolt holes.
S — Distance from centerline of pump to centerline of suction nozzle.
U — Diameter of straight shaft—coupling end.
V — Length of shaft available for coupling or pulley.
X — Distance from discharge face to centerline of pump.
Y — Horizontal distance—centerline discharge nozzle to suction nozzle face.
Z — Centerline discharge nozzle to centerline of pump.

Note: Where multiple dimensions for similar components are required, i.e., mounting pad widths and locations, subscripts 1, 2, 3, et cetera should be used. Number from right to left, i.e., HE_1, HE_2, HE_3. These subscript designations may appear in a view other than indicated.

FIGURE 4-29 Overhung impeller, separately coupled, single-stage, centerline-mounted pump. *(Courtesy: Hydraulics Institute.)*

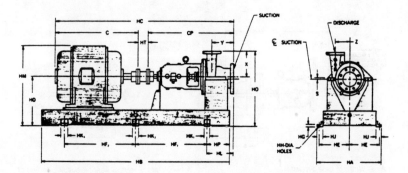

C —Length of driver.
CP —Length of pump.
HA —Width of base support.
HB —Length of base support.
HC —Overall length of combined pump and driver
    when on base.
HD —Vertical height—bottom of base support to
    centerline of pump.
HE —Distance from centerline pump to centerline
    hold-down bolts.
HF —Distance from centerline to centerline of hold-
    down bolt holes.
HG —Thickness of pads on support, or height of
    base plate, depending on location of bolt
    holes.
HH —Diameter of hold-down bolt holes.
HJ —Width of pads for hold-down bolts.
HK —Length of support pad for hold-down bolts.
HL —Horizonal distance from suction nozzle face
    to centerline nearest hold-down bolt holes.

HM —Height of unit—bottom of base to top of
    driver.
HO —Vertical distances—bottom of support to dis-
    charge nozzle face or top of case on hori-
    zontally split pumps.
HP —Length from edge of support, or base plate,
    to centerline of bolt holes.
HT —Horizontal distance—between pump and driv-
    ing shaft.
S —Distance from centerline of pump to center-
    line of suction nozzle.
X —Distance from discharge face to centerline
    of pump.
Y —Horizontal distance—centerline discharge
    nozzle to suction nozzle face.
Z —Centerline discharge nozzle to centerline of
    pump.

Note: Where multiple dimensions for similar components are required, i.e., mounting pad widths and locations,
subscripts 1, 2, 3, et cetera should be used. Number from right to left, i.e., HE_1, HE_2, HE_3. These subscript designations
may appear in a view other than indicated.

**FIGURE 4-30 Overhung impeller, separately coupled, single-stage, center-
line-mounted pump on baseplate.** *(Courtesy: Hydraulics Institute.)*

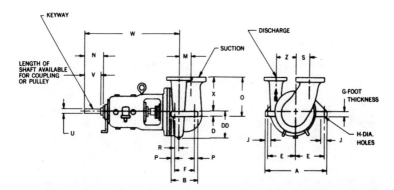

A —Width of base support.

B —Length of base support.

D —Vertical height—bottom of base support to centerline of pump.

DD —Distance—pump centerline to bottom drain plug.

E —Distance from centerline pump to centerline hold-down bolts.

F —Distance from centerline to centerline of hold-down bolt holes.

G —Thickness of pads on support, or height of base plate, depending on location of bolt holes.

H —Diameter of hold-down bolt holes.

J —Width of pads for hold-down bolts.

M —Horizontal distance from centerline of discharge flange to centerline of suction flange.

N —Distance—end of bearing housing to end of shaft.

O —Vertical distance—bottom of support to discharge nozzle face or top of case on horizontally split pumps.

P —Length from edge of support, or base plate, to centerline of bolt holes.

R —Horizontal distance—centerline discharge flange to centerline hold-down bolt hole.

S —Distance from centerline of pump to centerline of suction nozzle.

U —Diameter of straight shaft—coupling end.

V —Length of shaft available for coupling or pulley.

W —Distance from centerline of discharge flange to end of pump shaft.

X —Distance from discharge face to centerline of pump.

Z —Centerline discharge nozzle to centerline of pump.

Note: Where multiple dimensions for similar components are required, i.e., mounting pad widths and locations, subscripts 1, 2, 3, et cetera should be used. Number from right to left, i.e., HE_1, HE_2, HE_3. These subscript designations may appear in a view other than indicated.

FIGURE 4-31 Overhung impeller, separately coupled, single-stage, center-line-mounted (top-suction) pump. *(Courtesy: Hydraulics Institute.)*

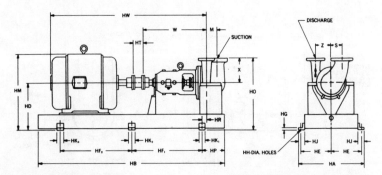

HA —Width of base support.

HB —Length of base support.

HD —Vertical height—bottom of base support to centerline of pump.

HE —Distance from centerline pump to centerline hold-down bolts.

HF —Distance from centerline to centerline of hold-down bolt holes.

HG —Thickness of pads on support, or height of base plate, depending on location of bolt holes.

HH —Diameter of hold-down bolt holes.

HJ —Width of pads for hold-down bolts.

HK —Length of support pad for hold-down bolts.

HM —Height of unit—bottom of base to top of driver.

HO —Vertical distance—bottom of support to discharge nozzle face or top of case on horizontally split pumps.

HP —Length from edge of support, or base plate, to centerline of bolt holes.

HR —Horizontal distance—centerline discharge flange to centerline hold-down bolt hole.

HT —Horizontal distance—between pump and driving shaft.

HW —Distance from centerline of discharge flange to end of motor.

M —Horizontal distance from centerline of discharge flange to centerline of suction flange.

S —Distance from centerline of pump to centerline of suction nozzle.

W —Distance from centerline of discharge flange to end of pump shaft.

X —Distance from discharge face to centerline of pump.

Z —Centerline discharge nozzle to centerline of pump.

Note: Where multiple dimensions for similar components are required, i.e., mounting pad widths and locations, subscripts 1, 2, 3, et cetera should be used. Number from right to left, i.e., HE_1, HE_2, HE_3. These subscript designations may appear in a view other than indicated.

FIGURE 4-32   Overhung impeller, separately coupled, single-stage, center-line-mounted pump on baseplate (top suction). *(Courtesy: Hydraulics Institute.)*

Vane

In this type, the vane or vanes, which may be in the form of blades, buckets, rollers, or slippers, co-operate with a cam to draw fluid into and force it from the pump chamber. These pumps may be made with vanes in either the rotor or stator and with radial hydraulic forces on the rotor balanced or unbalanced. The vane-in-rotor pumps may be made with constant or variable displacement pumping elements. Fig. 2 illustrates a vane-in-rotor constant displacement unbalanced pump. Fig. 3 illustrates a vane-in-stator constant displacement unbalanced pump.

Piston

In this type, fluid is drawn in and forced out by pistons which reciprocate within cylinders with the valving accomplished by rotation of the pistons and cylinders relative to the ports. The cylinders may be axially or radially disposed and arranged for either constant or variable displacement pumping. All types are made with multiple pistons except that the constant displacement radial type may be either single or multiple piston. Fig. 4 illustrates an axial, constant displacement piston pump.

Flexible Member

In this type, the fluid pumping and sealing action depends on the elasticity of the flexible member(s). The flexible member may be a tube, a vane, or a liner. These types are illustrated in Figs. 5, 6, and 7, respectively.

Lobe

In this type, fluid is carried between rotor lobe surfaces from the inlet to the outlet. The rotor surfaces cooperate to provide continuous sealing. The rotors must be timed by separate means. Each rotor has one or more lobes. Figs. 8 and 9 illustrate a single and a three-lobe pump, respectively.

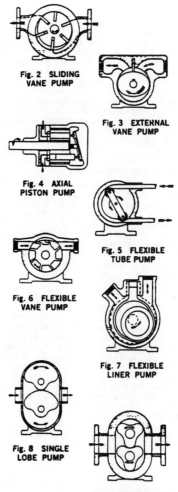

Fig. 2 SLIDING VANE PUMP

Fig. 3 EXTERNAL VANE PUMP

Fig. 4 AXIAL PISTON PUMP

Fig. 5 FLEXIBLE TUBE PUMP

Fig. 6 FLEXIBLE VANE PUMP

Fig. 7 FLEXIBLE LINER PUMP

Fig. 8 SINGLE LOBE PUMP

Fig. 9 THREE-LOBE PUMP

FIGURE 4-33 Basic rotary pumps. *(Courtesy: Hydraulics Institute.)*

Gear

In this type. fluid is carried between gear teeth and displaced when they mesh. The surfaces of the rotors cooperate to provide continuous sealing and either rotor is capable of driving the other.

External gear pumps have all gear rotors cut externally. These may have spur, helical, or herringbone gear teeth and may use timing gears.

Internal gear pumps have one rotor with internally cut gear teeth meshing with an externally cut gear. Pumps of this class are made with or without a crescent shaped partition. Fig. 10 illustrates an external spur gear pump. Figs. 11 and 12 illustrate internal gear pumps with and without the crescent shaped partition.

Circumferential Piston

In this type, fluid is carried from inlet to outlet in spaces between piston surfaces. There are no sealing contacts between rotor surfaces. In the external circumferential piston pump, the rotors must be timed by separate means, and each rotor may have one or more piston elements. In the internal circumferential piston pump, timing is not required, and each rotor must have two or more piston elements. Fig. 13 illustrates an external multiple piston type. The grey portions of the figure represent the rotating parts.

Screw

In this type. fluid is carried in spaces between screw threads and is displaced axially as they mesh.

Single screw pumps. (commonly called progressing cavity pumps) illustrated in Fig. 14, have a rotor with external threads and a stator with internal threads. The rotor threads are eccentric to the axis of rotation.

Screw and wheel pumps, shown in Fig. 15, depend upon a plate wheel to seal the screw so that there is no continuous cavity between the inlet and outlet.

Multiple screw pumps have multiple external screw threads. Such pumps may be timed or untimed. Fig. 16 illustrates a timed screw pump. Fig. 17 illustrates an untimed screw pump.

Fig. 10 EXTERNAL GEAR PUMP

Fig. 11 INTERNAL GEAR PUMP (with crescent)

Fig. 12 INTERNAL GEAR PUMP (without crescent)

Fig. 13 CIRCUMFERENTIAL PISTON PUMP

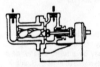
Fig. 14 SINGLE SCREW PUMP (progressing cavity)

Fig. 15 SCREW AND WHEEL PUMP

Fig. 16 TWO SCREW PUMP

Fig. 17 THREE SCREW PUMP

FIGURE 4-33 *(Continued)*

Purpose

The nomenclature and definitions in these Standards identify the various pump components and provide terminology which will be mutually understandable to the purchaser, to the manufacturer, and to anyone writing specifications for pumps and pumping equipment.

Fluids and Liquids

In these Standards, the term "fluid" covers liquids, gases, vapors, and mixtures thereof. The word "liquid" is used only to describe true liquids that are free of vapors and solids. The word "fluid" is more general and is used to describe liquids that may contain, or be mixed with, matter in other than the liquid phase.

Pumping Chamber

The pumping chamber is the space formed by the body and end plate(s), into which fluid is drawn and from which fluid is discharged by the action of the rotor(s).

Rotating Assembly

The rotating assembly generally consists of all rotating parts essential to the pumping action but also may include other parts specified by the manufacturer.

Body

The body is an external part which surrounds the periphery of the pumping chamber and which also may form one end plate. It is sometimes called a casing or a housing.

End Plate

An end plate is a part which closes an end of the body to form the pumping chamber. One or more are used, depending on the construction of the pump. It is sometimes called a head or cover.

Rotor

A rotor is a part which rotates in the pumping chamber. One or more are used per pump. It is sometimes referred to by a specific name such as gear, screw, impeller, etc.

Bearing

A bearing is a part which supports or positions the shafts on which a rotor is mounted. A bearing may be internal (wetted by the liquid being pumped) or external, and may be either an antifriction (ball or roller bearing) or fluid film (sleeve and journal) type.

Timing Gear

A timing gear is a part used to transmit torque from one rotor shaft to another, and to maintain the proper angular relationship of the rotors. It may be outside the pumping chamber and is sometimes called a pilot gear.

Relief Valve

A relief valve is a mechanism designed to control or to limit pressure by the opening of an auxiliary passage at a predetermined pressure.

A relief valve may be either integral with the body or end plate, or attachable. It may be adjustable through a predetermined range of pressures, or have a fixed setting. It may be designed to bypass the fluid internally from the pump outlet to the pump inlet, or externally through an auxiliary port.

Terms commonly used in specifying performance are:

Cracking pressure—sometimes called set pressure, start-to-discharge pressure, or popping pressure—the pressure at which the valve just starts to open. This pressure cannot be determined readily in a valve which bypasses the liquid within the pump.

Full-flow bypass pressure—the pressure at which the full output of the pump flows through the valve and the auxiliary passage.

Reseating pressure—the pressure at which the valve closes completely. This pressure is usually below the cracking pressure and is difficult to measure accurately when the liquid is bypassed within the pump.

Per cent overpressure, sometimes called per cent accumulation or per cent regulation—the difference between bypass pressure and cracking pressure, expressed as per cent of cracking pressure.

Stuffing Box

A stuffing box is a cylindrical cavity through which a shaft extends and in which leakage at the shaft is controlled by means of packing and a gland.

Gland

A gland is a part which may be adjusted to compress packing in a stuffing box. It is sometimes called a gland follower.

FIGURE 4-34   Basic components of rotary pumps. *(Courtesy: Hydraulics Institute.)*

Inlet or Suction Port

One or more openings in the pump through which the pumped fluid may enter the pumping chamber.

Outlet or Discharge Port

One or more openings in the pump through which the pumped fluid may leave the pumping chamber.

Lantern Ring

A lantern ring is an annular ring located in a stuffing box to provide space between or adjacent to packing rings for the introduction of a lubricant or a barrier fluid, the circulation of a cooling medium, or the relief of pressure against the packing. It is sometimes called a seal cage.

Seal Chamber

A seal chamber is a cavity through which a shaft extends and in which leakage at the shaft is controlled by means of a mechanical seal or a radial seal.

Mechanical Seal

A mechanical seal is a device located in a seal chamber and consists of rotating and stationary elements with opposed seal faces. A rotating element is fastened and sealed to the shaft. A stationary element is mounted and sealed to the end plate or body. At least one element is loaded in an axial direction so that the seal faces of the elements are maintained in close proximity to each other at all times. Usually, the seal faces are highly lapped surfaces on materials selected for low friction and for resistance

FIGURE 4-34   *(Continued)*

to corrosion by the fluids to be pumped. Mechanical seals are sometimes called face type seals.

Radial Seal

A radial seal is a device located in a seal chamber which seals on its outside diameter through an interference fit with its mating bore, and on the rotating shaft with a flexible, radially loaded surface. Radial seals include lip type seals, O rings, V cups, U cups, etc., and may or may not be spring loaded.

Direction of Rotation

Drive shaft rotation is designated as "clockwise" (CW) or "counterclockwise" (CCW) as determined when viewing the pump from the driver end.

Jacketed Pump

A jacketed pump is one in which the body and/or end plates incorporate passageways through which steam, oil, water, or other fluid can be circulated to control the temperature of the pump or the fluid in the pump.

Letter (Dimensional) Designations

The letter designations used on the following drawings were prepared to provide a common means for identifying various pump dimensions and also to serve as a common language which will be mutually understandable to the purchaser, manufacturer and to anyone writing specifications for pumps and pumping equipment.

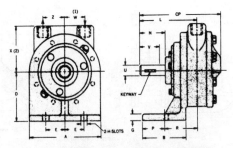

INTERNAL GEAR PUMP—FOOT MOUNTING

[1]If equal to "Z", use "Z". These do not change for horizontal nozzles.
[2]If not equal for suction and discharge, use Y for suction and show separately.

A —Width of base support or foot, or width of pump.
AF —Mounting hole bolt circle.
AJ —Diameter, mounting pilot fit.
B —Length of base support or foot.
CP —Length of pump.
D —Vertical height—centerline of pump shaft to bottom of base support or foot.
E —Distance—centerline of pump shaft to centerline of hold-down bolt hole or slot.
G —Thickness of pads on support, or height of base plate, depending on location of bolt holes or slots.
H —Diameter of hold-down or mounting bolt holes or slots.
L —Distance—centerline of outlet port to end of pump shaft.

LP —Length of adapter piece or length of pilot fit.
N —Distance—end of shaft to nearest obstruction.
P —Length from edge of support, or base plate to centerline of bolt hole.
R —Horizontal distance—centerline outlet port to centerline of hold-down bolt hole.
S —Distance—end of pump shaft to face of mounting flange.
U —Diameter of straight shaft—coupling end.
V —Length of shaft available for coupling or pulley.
W —Distance—centerline of shaft to centerline of inlet port.
X —Distance—centerline of pump shaft to outlet port.
Z —Distance—centerline of pump shaft to centerline of ports.

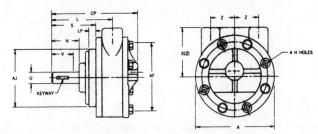

INTERNAL GEAR PUMP—FLANGE MOUNTING

FIGURE 4-35 Internal gear pump—foot mounting and flange mounting.
(Courtesy: Hydraulics Institute.)

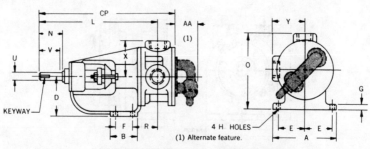

INTERNAL GEAR PUMP—FRAME MOUNTING

A —Width of base support or foot, or width of pump.
AA —Added length for alternate feature.
B —Length of base support or foot.
CP —Length of pump.
D —Vertical height—centerline of pump shaft to bottom of base support or foot.
E —Distance—centerline of pump shaft to centerline of hold-down bolt hole.
F —Distance—centerline to centerline of hold-down bolt holes.
G —Thickness of pads on support, or height of base plate, depending on location of bolt holes.
H —Diameter of hold-down or mounting bolt holes.
J —Width of pads for hold-down bolts.
K —Length of pads for hold-down bolts.
L —Distance—centerline of outlet port to end of pump shaft.
LP —Length of adapter piece or length of pilot fit.

MP —Distance—centerline of outlet port to mounting flange of adapter.
N —Distance—end of shaft to nearest obstruction.
O —Vertical distance—bottom of support to outlet port or top of pump.
OP —Vertical distance—bottom of support to outlet port or top of pump.
R —Horizontal distance—centerline outlet port to centerline of hold-down bolt hole.
U —Diameter of straight shaft—coupling end.
V —Length of shaft available for coupling or pulley.
X —Distance—centerline of pump shaft to outlet port.
XR —Distance—centerline of motor hold-down bolt to centerline of conduit box.
Y —Distance—centerline of pump shaft to inlet port.
Z —Distance—centerline of pump shaft to centerline of ports.

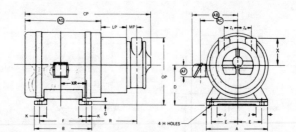

Note: Circled letters are NEMA designations. If Z_1 and Z_2 are not equal, substitute W_1 for Z_1, the distance from centerline of pump to centerline of inlet.

INTERNAL GEAR PUMP—CLOSE COUPLED

FIGURE 4-36 Internal gear pump—frame mounting and close-coupled.
(Courtesy: Hydraulics Institute.)

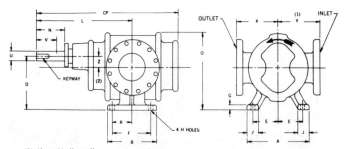

(1) If equal to X, use X.
(2) If not equal for inlet and outlet use W for inlet and show both in end view.

EXTERNAL GEAR PUMP—FLANGED PORTS

A —Width of base support or foot, or width of pump.
B —Length of base support or foot.
CP —Length of pump.
D —Vertical height—centerline of pump shaft to bottom of base support or foot.
E —Distance—centerline of pump shaft to centerline of hold-down bolt hole.
F —Distance—centerline to centerline of hold-down bolt holes.
G —Thickness of pads on support, or height of base plate, depending on location of bolt holes.
H —Diameter of hold-down or mounting bolt holes.
J —Width of pads for hold-down bolts.
L —Distance—centerline of outlet port to end of pump shaft.

N —Distance—end of shaft to nearest obstruction.
O —Vertical distance—bottom of support to outlet port or top of pump.
R —Horizontal distance—centerline outlet port to centerline of hold-down bolt hole.
U —Diameter of straight shaft—coupling end.
V —Length of shaft available for coupling or pulley.
W —Distance—centerline of shaft to centerline of inlet port.
X —Distance—centerline of pump shaft to outlet port.
Y —Distance—centerline of pump shaft to inlet port.
Z —Distance—centerline of pump shaft to centerline of ports.

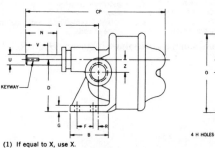

(1) If equal to X, use X.

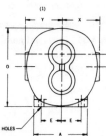

EXTERNAL GEAR PUMP—THREADED PORTS

FIGURE 4-37 External gear pump—flanged ports and threaded ports. *(Courtesy: Hydraulics Institute.)*

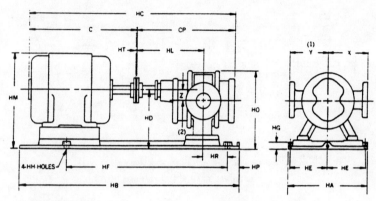

(1) If equal to X, use X.
(2) If not equal for inlet and outlet use W for inlet and show both in end view.

C —Length of driver.
CP —Length of pump.
HA —Width of base support or foot, or width of pump.
HB —Length of base support or foot.
HC —Overall length of combined pump and driver when on base.
HD —Vertical height—centerline of pump shaft to bottom of base support or foot.
HE —Distance—centerline of pump shaft to centerline of hold-down bolt hole.
HF —Distance—centerline to centerline of hold-down bolt holes.
HG —Thickness of pads on support, or height of base plate, depending on location of bolt holes.
HH —Diameter of hold-down or mounting bolt holes.

HL —Distance—centerline of outlet port to end of pump shaft.
HM —Height of unit—bottom of base to top of driver.
HO —Vertical distance—bottom of support to outlet port or top of pump.
HP —Length from edge of support, or base plate to centerline of bolt hole.
HR —Horizontal distance—centerline outlet port to centerline of hold-down bolt hole.
HT —Horizontal distance—between pump and driving shaft.
X —Distance—centerline of pump shaft to outlet port.
Y —Distance—centerline of pump shaft to inlet port.
Z —Distance—centerline of pump shaft to centerline of ports.

FIGURE 4-38   External gear pump on baseplate. *(Courtesy: Hydraulics Institute.)*

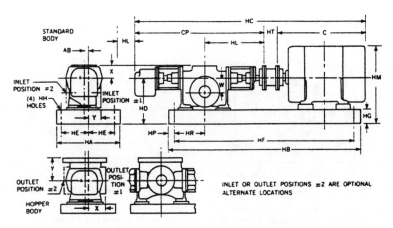

AB —Distance—pump centerline to pump shaft centerline.
C —Length of driver.
CP —Length of pump.
HA —Width of base support or foot or width of pump.
HB —Length of base support or foot.
HC —Overall length of combined pump and driver when on base.
HD —Vertical height—centerline of pump shaft to bottom of base or foot.
HE —Distance—centerline of pump shaft to centerline of hold-down bolt.
HF —Distance—centerline to centerline of hold-down bolt hole.
HG —Baseplate height above foundation.
HH —Diameter of hold-down or mounting bolt holes.

HL —Distance—centerline of outlet port to end of pump shaft.
HL'—Length required for removal of rotor (HL + ½ body length).
HM —Maximum height of unit from bottom of base.
HP —Length from edge of support or base plate to centerline of bolt hole.
HR —Horizontal distance—centerline outlet port to centerline of hold-down bolt.
HT —Horizontal distance—between pump shaft and driving shaft.
W —Distance—centerline of pump shaft to centerline of inlet port.
X —Distance—centerline of pump shaft to outlet port.
Y —Distance—centerline of pump shaft to inlet port.

FIGURE 4-39   External gear and bearing screw pump on baseplate. *(Courtesy: Hydraulics Institute.)*

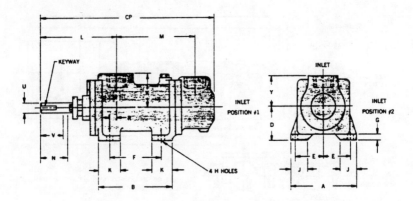

For inlet position 1 & 2, Y would be the distance from vertical centerline to face of inlet.

A —Width of base support or foot, or width of pump.
B —Length of base support or foot.
CP —Length of pump.
D —Vertical height—centerline of pump shaft to bottom of base support or foot.
E —Distance—centerline of pump shaft to centerline of hold-down bolt hole.
F —Distance—centerline to centerline of hold-down bolt holes.
G —Thickness of pads on support, or height of base plate, depending on location of bolt holes.
H —Diameter of hold-down or mounting bolt holes.
J —Width of pads for hold-down bolts.

K —Length of pads for hold-down bolts.
L —Distance—centerline of outlet port to end of pump shaft.
M —Horizontal distance from centerline to centerline of ports.
N —Distance—end of shaft to nearest obstruction.
R —Horizontal distance—centerline outlet port to centerline of hold-down bolt hole.
U —Diameter of straight shaft—coupling end.
V —Length of shaft available for coupling or pulley.
X —Distance—centerline of pump shaft to outlet port.
Y —Distance—centerline of pump shaft to inlet port.

FIGURE 4-40   Multiple-screw pump. *(Courtesy: Hydraulics Institute.)*

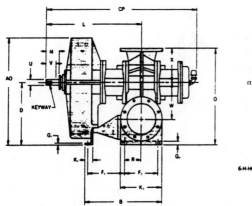
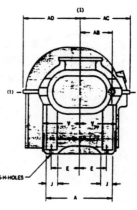

(1) Use these centerlines instead of centerline of pump shaft, since some models have alternate shaft positions.

A   —Width of base support or foot, or width of pump.

AB   —Distance—pump centerline to shaft centerline.

AC   —Distance—pump centerline to left side of gear box or other structure.

AD   —Distance—pump centerline to right side of gear box or other structure.

AO   —Height, overall, to top of gear or other structure.

B   —Length of base support or foot.

CP   —Length of pump.

D   —Vertical height—centerline of pump shaft to bottom of base support or foot.

E   —Distance—Centerline of pump to centerline of hold-down bolt hole.

F   —Distance—centerline to centerline of hold-down bolt holes.

G   —Thickness of pads on support, or height of base plate, depending on location of bolt holes.

H   —Diameter of hold-down or mounting bolt holes.

J   —Width of pads for hold-down bolts.

K   —Length of pads for hold-down bolts.

L   —Distance—centerline of outlet port to end of pump shaft.

N   —Distance—end of shaft to nearest obstruction.

O   —Vertical distance—bottom of support to outlet port or top of pump.

R   —Horizontal distance—centerline outlet port to centerline of hold-down bolt hole.

U   —Diameter of straight shaft—coupling end.

V   —Length of shaft available for coupling or pulley.

W   —Distance—Centerline of pump to centerline of inlet port.

X   —Distance—centerline of pump to outlet port.

Y   —Distance—centerline of pump to inlet port.

FIGURE 4-41   Lobe pump. *(Courtesy: Hydraulics Institute.)*

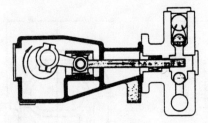

FIGURE 4-42 Horizontal single-acting plunger power pump. *(Courtesy: Hydraulics Institute.)*

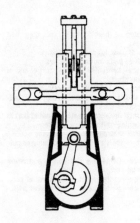

FIGURE 4-43 Vertical single-acting plunger power pump. *(Courtesy: Hydraulics Institute.)*

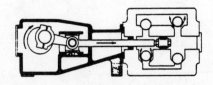

FIGURE 4-44 Horizontal double-acting piston power pump. *(Courtesy: Hydraulics Institute.)*

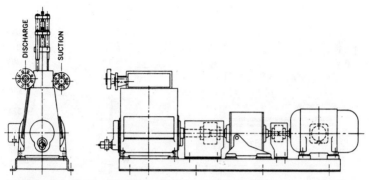

FIGURE 4-45 Vertical triplex plunger pump, on base, gear reduction. *(Courtesy: Hydraulics Institute.)*

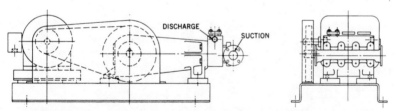

FIGURE 4-46 Horizontal triplex plunger pump, on base, belt drive. *(Courtesy: Hydraulics Institute.)*

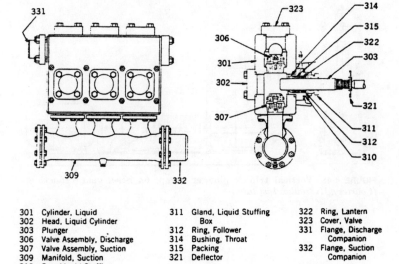

301	Cylinder, Liquid	311	Gland, Liquid Stuffing	322	Ring, Lantern
302	Head, Liquid Cylinder		Box	323	Cover, Valve
303	Plunger	312	Ring, Follower	331	Flange, Discharge
306	Valve Assembly, Discharge	314	Bushing, Throat		Companion
307	Valve Assembly, Suction	315	Packing	332	Flange, Suction
309	Manifold, Suction	321	Deflector		Companion
310	Box, Liquid Stuffing				

FIGURE 4-47 **Liquid end, horizontal plunger power pump.** *(Courtesy: Hydraulics Institute.)*

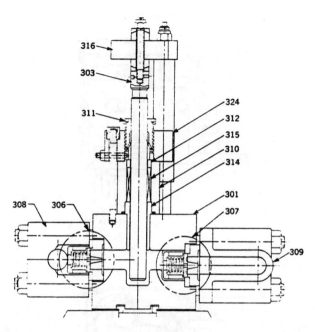

301	Cylinder, Liquid	309	Manifold, Suction
303	Plunger	310	Box, Liquid Stuffing
306	Valve Assembly, Discharge	311	Gland, Liquid Stuffing
307	Valve Assembly, Suction		Box
308	Manifold, Discharge	312	Ring, Follower

314	Bushing, Throat
315	Packing
316	Crosshead, Upper
324	Ring, Gland

FIGURE 4-48   Liquid end, vertical plunger power pump. *(Courtesy: Hydraulics Institute.)*

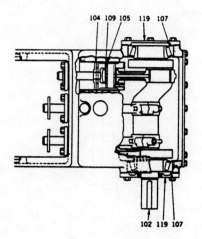

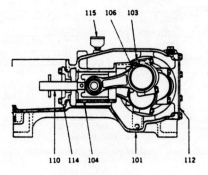

101	Frame, Power	106	Bearing, Crankpin	112	Cover, Crankcase
102	Crankshaft	107	Bearing, Main	114	Box, Wiper
103	Rod, Connecting		Crankshaft	115	Breather
104	Crosshead, Power	109	Bearing, Wrist Pin	119	Housing, Crankshaft
105	Pin, Wrist	110	Extension, Crosshead		Bearing

FIGURE 4-49 Power end, horizontal plunger power pump. *(Courtesy: Hydraulics Institute.)*

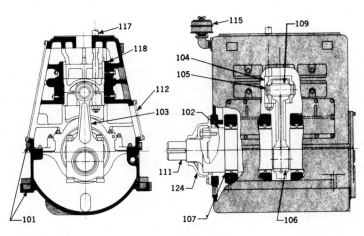

101	Frame, Power	107	Bearing, Main,	115	Breather
102	Crankshaft		Crankshaft	117	Rod, Side
103	Rod, Connecting	109	Bearing, Wrist Pin	118	Way, Crosshead
104	Crosshead, Power	111	Extension Crankshaft	124	Cover, Crankshaft
105	Pin, Wrist	112	Cover, Crankcase		Extension
106	Bearing, Crankpin				

FIGURE 4-50 **Power end, vertical plunger power pump.** *(Courtesy: Hydraulics Institute.)*

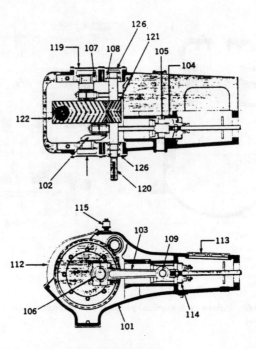

101	Frame, Power	108	Bearing, Pinion Shaft	119	Housing, Bearing Crankshaft
102	Crankshaft	109	Bearing, Wrist Pin	120	Pinion Shaft
103	Rod, Connecting	112	Cover, Crankcase	121	Pinion
104	Crosshead, Power	113	Cover, Cradle	122	Gear
105	Pin, Wrist	114	Box, Wiper	126	Housing, Bearing, Pinion Shaft
106	Bearing, Crankpin	115	Breather		
107	Bearing, Main, Crankshaft				

FIGURE 4-51 **Power end, horizontal duplex power pump with integral gears.** *(Courtesy: Hydraulics Institute.)*

TABLE 4-2   Summary of Operating Performances of Pumps

Pump Type/Style	Solids Toler- ance	Capacity (dm³/s)	(gph)	Max. Head (m)	(ft)
Centrifugal					
Horizontal					
Single-stage overhung	MH	1 ~ 320	950 ~ 3×10^5	150	492
2-stage overhung	MH	1 ~ 75	950 ~ 7.1×10^4	425	1394
Single-stage impeller between bearings	MH	1 ~ 2500	950 ~ 2.4×10^6	335	1099
Chemical	MH	65	6.2×10^4	73	239
Slurry	H	65	6.2×10^4	120	394
Canned	L	0.1 ~ 1250	95 ~ 1.2×10^6	1500	4922
Multi. horiz. split	M	1 ~ 700	950 ~ 6.7×10^5	1675	5495
Multi., barrel type	M	1 ~ 550	950 ~ 5.2×10^5	1675	5495
Vertical					
Single-stage process	M	1 ~ 650	950 ~ 6.2×10^5	245	804
Multistage	M	1 ~ 5000	950 ~ 4.8×10^6	1830	6004
In-line	M	1-750	950 ~ 7.1×10^5	215	705
High-speed	L	0.3 ~ 25	285 ~ 2.4×10^4	1770	5807
Sump	MH	1 ~ 45	950 ~ 4.3×10^4	60	197
Multi. deep well	M	0.3 ~ 25	285 ~ 2.4×10^4	1830	6004
Axial (propeller)	H	1 ~ 6500	950 ~ 6.2×10^6	12	39.4
Turbine (regenerative)	M	0.1 ~ 125	95 ~ 1.2×10^5	760	2493
Positive Displacement				(kPa)	(psi)
Reciprocating					
Piston, plunger	M	1 ~ 650	950 ~ 6.2×10^5	345000	50038
Metering	L	0 ~ 1	0 ~ 950	517000	74985
Diaphragm	L	0.1 ~ 6	95 ~ 5.7×10^3	34500	5004
Rotary					
Screw	M	0.1 ~ 125	95 ~ 1.2×10^5	20700	3002
Gear	M	0.1 ~ 320	95 ~ 3.0×10^5	3400	493

MH = moderately high, H = high, M = medium, L = low.

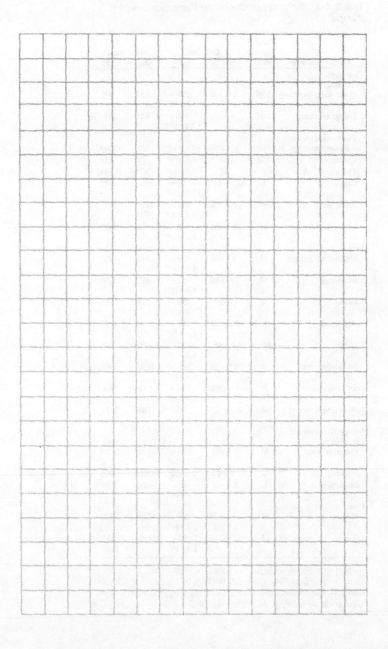

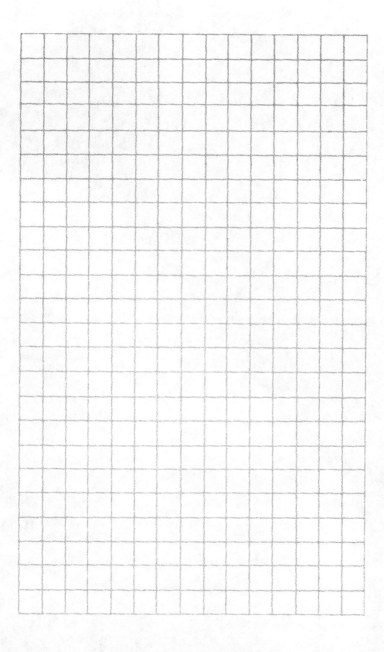

Weirs, Flumes, and Orifices

TABLE 5-1 Discharge from Triangular Notch Weirs with End Contractions

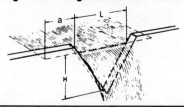

Head (H) in Feet	Flow in Gallons Per Min. 90° Notch	Flow in Gallons Per Min. 60° Notch	Head (H) in Feet	Flow in Gallons Per Min. 90° Notch	Flow in Gallons Per Min. 60° Notch	Head (H) in Feet	Flow in Gallons Per Min. 90° Notch	Flow in Gallons Per Min. 60° Notch
0.02	0.06	0.04	0.72	481	278	1.42	2627	1518
0.04	0.35	0.20	0.74	515	298	1.44	2721	1573
0.06	0.96	0.56	0.76	551	318	1.46	2816	1628
0.08	1.98	1.14	0.78	587	340	1.48	2913	1684
0.10	3.46	2.00	0.80	626	362	1.50	3013	1741
0.12	5.45	3.15	0.82	666	385	1.52	3114	1800
0.14	8.02	4.63	0.84	707	409	1.54	3218	1860
0.16	11.20	6.47	0.86	750	433	1.56	3323	1921
0.18	15.03	8.69	0.88	794	459	1.58	3431	1983
0.20	19.56	11.30	0.90	840	486	1.60	3540	2046
0.22	24.8	14.3	0.92	888	513	1.62	3652	2111
0.24	30.9	17.8	0.94	937	541	1.64	3766	2177
0.26	37.7	21.8	0.96	987	571	1.66	3882	2244
0.28	45.4	26.2	0.98	1040	601	1.68	4000	2312
0.30	53.9	31.2	1.00	1093	632	1.70	4120	2381
0.32	63.3	36.6	1.02	1149	664	1.72	4242	2452
0.34	73.7	42.6	1.04	1206	697	1.74	4366	2524
0.36	85.0	49.1	1.06	1265	731	1.76	4493	2597
0.38	97.3	56.3	1.08	1325	766	1.78	4622	2671
0.40	111	63.9	1.10	1388	802	1.80	4753	2747
0.42	125	72	1.12	1451	839	1.82	4886	2824
0.44	140	81	1.14	1517	877	1.84	5021	2902
0.46	157	91	1.16	1585	916	1.86	5159	2982
0.48	175	101	1.18	1654	956	1.88	5299	3063
0.50	193	112	1.20	1725	997	1.90	5441	3145
0.52	213	123	1.22	1797	1039	1.92	5585	3228
0.54	234	135	1.24	1872	1082	1.94	5731	3313
0.56	257	148	1.26	1948	1126	1.96	5880	3399
0.58	280	162	1.28	2027	1171	1.98	6031	3486
0.60	305	176	1.30	2107	1218	2.00	6185	3575
0.62	331	191	1.32	2189	1265			
0.64	358	207	1.34	2273	1314			
0.66	387	224	1.36	2358	1363			
0.68	417	241	1.38	2446	1414			
0.70	448	259	1.40	2536	1466			

Based on equation:

$$Q = (C)(4/15)(L)(H)\sqrt{2gH}$$

in which Q = flow of water in cu. ft. per sec.

L = width of notch in ft. at H distance above apex.

H = head of water above apex of notch in ft.

C = constant varying with conditions, .57 being used for this table.

a = should be not less than 3/4 L.

source: National Clay Pipe Institute.

For 90° notch the equation becomes

$$Q = 2.436H^{5/2}$$

For 60° notch the equation becomes

$$Q = 1.408H^{5/2}$$

TABLE 5-2 Discharge from Rectangular Weir with End Contractions

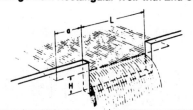

Figures in Table are in Gallons Per Minute

Head (H) in Inches	Length (L) of Weir in Feet 1	3	5	Additional g.p.m. for each ft. over 5 ft.	Head (H) in Inches	Length (L) of Weir in Feet 3	5	Additional g.p.m. for each ft. over 5 ft.
¼	4.5	13.4	22.4	4.5	7¾	2,238	3,785	774
½	12.8	38.2	63.8	12.8	8	2,338	3,956	814
¾	23.4	70.2	117.0	23.4	8¼	2,442	4,140	850
1	35.4	108	180	36.1	8½	2,540	4,312	890
1¼	49.5	150	250	50.4	8¾	2,656	4,511	929
1½	64.9	197	330	66.2	9	2,765	4,699	970
1¾	81.0	248	415	83.5	9¼	2,876	4,899	1,011
2	98.5	302	506	102	9½	2,985	5,098	1,051
2¼	117	361	605	122	9¾	3,101	5,288	1,091
2½	136	422	706	143	10	3,216	5,490	1,136
2¾	157	485	815	165	10½	3,480	5,940	1,230
3	178	552	926	187	11	3,716	6,355	1,320
3¼	200	624	1,047	211	11½	3,960	6,780	1,410
3½	222	695	1,167	236	12	4,185	7,165	1,495
3¾	245	769	1,292	261	12½	4,430	7,595	1,575
4	269	846	1,424	288	13	4,660	8,010	1,660
4¼	294	925	1,559	316	13½	4,950	8,510	1,780
4½	318	1,006	1,696	345	14	5,215	8,980	1,885
4¾	344	1,091	1,835	374	14½	5,475	9,440	1,985
5	370	1,175	1,985	405	15	5,740	9,920	2,090
5¼	396	1,262	2,130	434	15½	6,015	10,400	2,165
5½	422	1,352	2,282	465	16	6,290	10,900	2,300
5¾	449	1,442	2,440	495	16½	6,565	11,380	2,410
6	477	1,535	2,600	538	17	6,925	11,970	2,520
6¼		1,632	2,760	560	17½	7,140	12,410	2,640
6½		1,742	2,920	596	18	7,410	12,900	2,745
6¾		1,826	3,094	630	18½	7,695	13,410	2,855
7		1,928	3,260	668	19	7,980	13,940	2,970
7¼		2,029	3,436	702	19½	8,280	14,460	3,090
7½		2,130	3,609	736				

$$Q = 3.33 \ (L - 0.2H) \ H^{1.5}$$

in which

Q = cu. ft. of water flowing per second.

L = length of weir opening in feet (should be 4 to 8 times H).

H = head on weir in feet (to be measured at least 6 ft. back of weir opening).

a = should be at least $3H$.

SOURCE: National Clay Pipe Institute.

TABLE 5-3   Minimum and Maximum Recommended Flow
Rates for Rectangular Weirs (*a*) with End Contractions
and (*b*) without End Contractions

CREST LENGTH, FT.	MIN. HEAD, FT.	MIN. FLOW RATE		MAX. HEAD, FT.	MAX. FLOW RATE	
		MGD	CFS		MGD	CFS
1	0.2	.185	.286	0.5	.685	1.06
1½	0.2	.281	.435	0.75	1.89	2.92
2	0.2	.377	.584	1.0	3.87	5.99
2½	0.2	.474	.733	1.25	6.77	10.5
3	0.2	.570	.882	1.5	10.7	16.5
4	0.2	.762	1.18	2.0	21.9	33.9
5	0.2	.955	1.48	2.5	38.3	59.2
6	0.2	1.15	1.77	3.0	60.4	93.4
8	0.2	1.53	2.37	4.0	124	192
10	0.2	1.92	2.97	5.0	217	335

(a)

CREST LENGTH, FT.	MIN. HEAD, FT.	MIN. FLOW RATE		MAX. HEAD, FT.	MAX. FLOW RATE	
		MGD	CFS		MGD	CFS
1	0.2	.192	.298	0.5	.761	1.18
1½	0.2	.289	.447	0.75	2.10	3.24
2	0.2	.385	.596	1.0	4.30	6.66
2½	0.2	.481	.745	1.25	7.52	11.6
3	0.2	.577	.894	1.5	11.9	18.4
4	0.2	.770	1.19	2.0	24.3	37.7
5	0.2	.962	1.49	2.5	42.5	65.8
6	0.2	1.16	1.79	3.0	67.1	104
8	0.2	1.54	2.38	4.0	138	213
10	0.2	1.92	2.98	5.0	241	372

(b)

SOURCE: Isco, Inc.

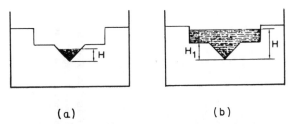

(a) (b)

FIGURE 5-1   Compound weir (90° V-notch weir with contracted rectangular weir). (a) Low flow—acts as V-notch weir only. (b) High flow—acts as combination of V-notch and rectangular weirs. (*Courtesy: Isco, Inc.*)

FIGURE 5-2   Inexpensive weir installation for small-stream measurement. (*Courtesy: Leupold & Stevens, Inc.*)

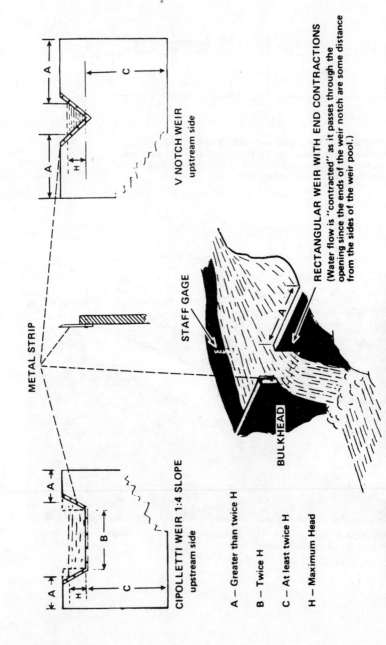

V NOTCH WEIR
upstream side

METAL STRIP

STAFF GAGE

BULKHEAD

RECTANGULAR WEIR WITH END CONTRACTIONS
(Water flow is "contracted" as it passes through the opening since the ends of the weir notch are some distance from the sides of the weir pool.)

CIPOLLETTI WEIR 1:4 SLOPE
upstream side

A – Greater than twice H

B – Twice H

C – At least twice H

H – Maximum Head

FIGURE 5-3 Typical sharp-crested weirs. (*Courtesy: Leupold & Stevens, Inc.*)

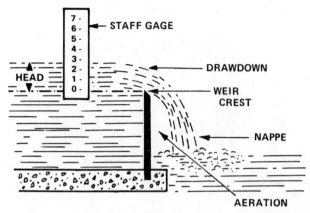

FIGURE 5-4 Sharp-crested weir with staff gauge. The staff gauge is located so that "0" on the gauge is the same elevation as the weir crest. (*Courtesy: Leupold & Stevens, Inc.*)

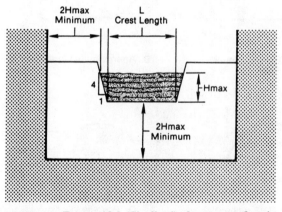

FIGURE 5-5 Trapezoidal (Cipolletti) sharp-crested weir. (*Courtesy: Isco, Inc.*)

TABLE 5-4 Minimum and Maximum Recommended Flow Rates for Cipolletti Weirs

CREST LENGTH, FT.	MIN. HEAD, FT.	MIN. FLOW RATE		MAX. HEAD, FT.	MAX. FLOW RATE	
		MGD	CFS		MGD	CFS
1	0.2	.195	.301	0.5	.769	1.19
1½	0.2	.292	.452	0.75	2.12	3.28
2	0.2	.389	.602	1.0	4.35	6.73
2½	0.2	.487	.753	1.25	7.60	11.8
3	0.2	.584	.903	1.5	12.0	18.6
4	0.2	.778	1.20	2.0	24.6	38.1
5	0.2	.973	1.51	2.5	43.0	66.5
6	0.2	1.17	1.81	3.0	67.8	105
8	0.2	1.56	2.41	4.0	139	214
10	0.2	1.95	3.01	5.0	243	375

SOURCE: Isco, Inc.

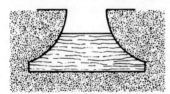

A. Sutro Or Proportional Weir

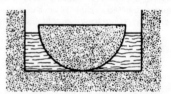

B. Approximate Linear Weir

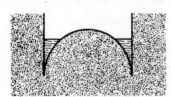

C. Approximate Exponential Weir

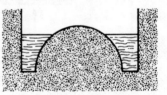

D. Poebing Weir

FIGURE 5-6 **Various other sharp-crested weir profiles.** (*Courtesy: Isco, Inc.*)

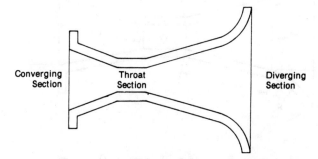

FIGURE 5-7   General flume configuration. (*Courtesy: Isco, Inc.*)

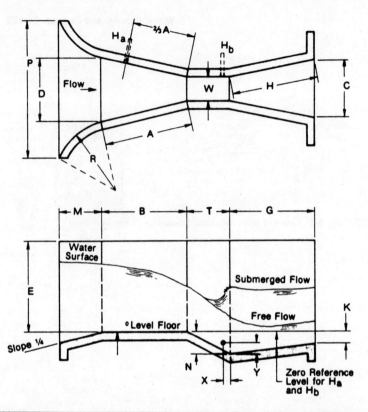

FIGURE 5-8 Parshall flume design with dimensions for various throat widths. (*Courtesy: Isco, Inc.*)

TABLE 5-5 Minimum and Maximum Recommended Flow
Rates for Free Flow-through Parshall Flumes

THROAT WIDTH, W	MIN. HEAD, FT.	MIN. FLOW RATE		MAX. HEAD, FT.	MAX. FLOW RATE	
		MGD	CFS		MGD	CFS
1 in.	0.07	.003	.005	0.60	.099	.153
2 in.	0.07	.007	.011	0.60	.198	.306
3 in.	0.10	.018	.028	1.5	1.20	1.86
6 in.	0.10	.035	.054	1.5	2.53	3.91
9 in.	0.10	.059	.091	2.0	5.73	8.87
1 ft.	0.10	.078	.120	2.5	10.4	16.1
1½ft.	0.10	.112	.174	2.5	15.9	24.6
2 ft.	0.15	.273	.423	2.5	21.4	33.1
3 ft.	0.15	.397	.615	2.5	32.6	50.4
4 ft.	0.20	.816	1.26	2.5	43.9	67.9
5 ft.	0.20	1.00	1.55	2.5	55.3	85.6
6 ft.	0.25	1.70	2.63	2.5	66.9	103
8 ft.	0.25	2.23	3.45	2.5	90.1	139
10 ft.	0.30	3.71	5.74	3.5	189	292
12 ft.	0.33	5.13	7.93	4.5	335	519

SOURCE: Isco, Inc.

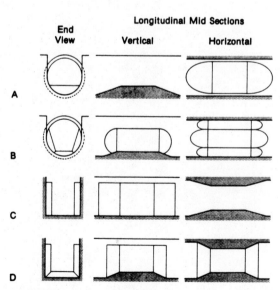

**FIGURE 5-9 Various cross-sectional shapes of Palmer-
Bowlus flumes.** (*Courtesy: Isco, Inc.*)

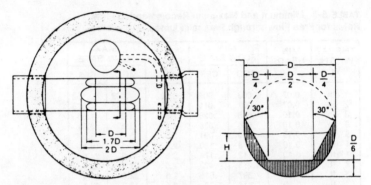

FIGURE 5-10 Dimensional configuration of standardized Palmer-Bowlus flume trapezoidal throat cross section. D = conduit diameter. (*Courtesy: Isco, Inc.*)

Depth D, ft.	Max. Capacity cfs
0.4	0.085
0.5	0.14
0.6	0.23
0.8	0.47
1.0	0.82

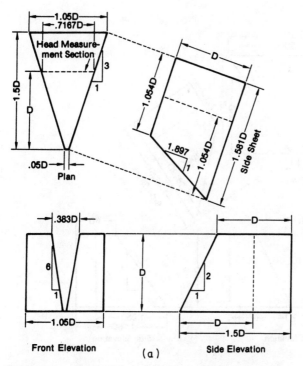

FIGURE 5-11 Dimensions and capacities of H-type flumes.
(a) HS flumes, (b) H flumes, (c) HL flumes. (*Courtesy: Isco, Inc.*)

Depth D, ft.	Max. Capacity cfs
0.5	0.35
0.75	0.97
1.0	1.99
1.5	5.49
2.0	11.3
2.5	19.7
3.0	31.1

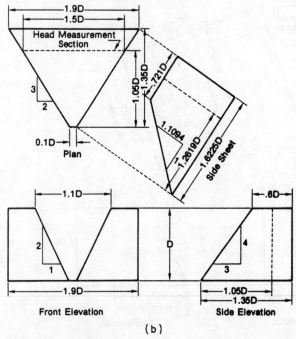

FIGURE 5-11 (*Continued*)

Depth D, ft.	Max. Capacity cfs
2.0	20.7
2.5	36.2
3.0	57.0
3.5	83.9
4.0	117.0

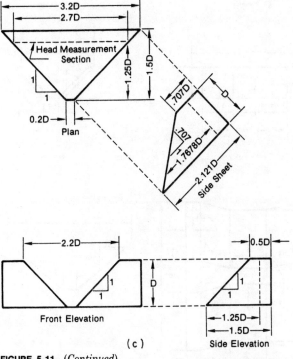

FIGURE 5-11 (*Continued*)

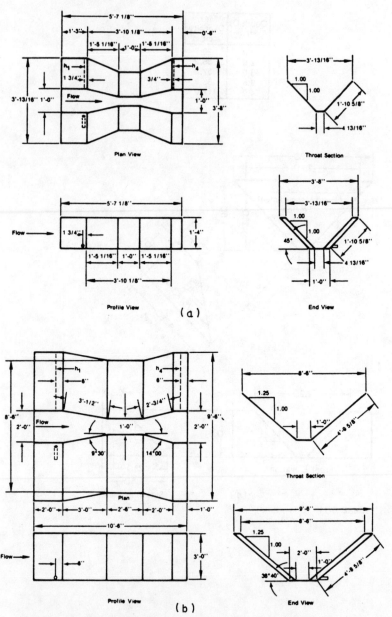

FIGURE 5-12 Trapezoidal flumes for 1- and 2-foot irrigation channels.
(*a*) For 1-foot channel, (*b*) for 2-foot channel. (*Courtesy: Isco, Inc.*)

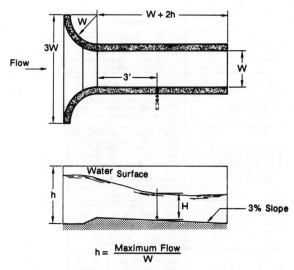

$$h = \frac{\text{Maximum Flow}}{W}$$

FIGURE 5-13 Original San Dimas flume. (*Courtesy: Isco, Inc.*)

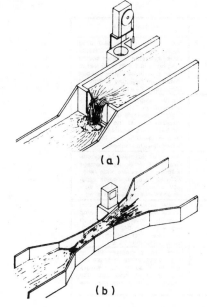

FIGURE 5-14 Weir versus flume typical installation as measuring devices. (*a*) Weir, (*b*) flume. (*Courtesy: Isco, Inc.*)

TABLE 5-6 Selection of a Primary Measuring Device: Weirs (*a*) versus Flumes (*b*)

PRIMARY MEASURING DEVICE		
GENERAL TYPE	**SPECIFIC TYPE**	**COMMENTS**
Advantages 1. Low cost. 2. Easy to install. **Disadvantages** 1. Fairly high head loss. 2. Must be periodically cleaned-not suitable for channels carrying excessive solids. 3. Accuracy affected by excessive approach velocities.	A. TRIANGULAR (V-NOTCH) WEIR	Accurate device particularly suited to measuring low flows. Best weir profile for discharges of less than 1 cfs and may be used for flows up to 10 cfs.
	B. RECTANGULAR (CONTRACTED) WEIR WITH END CONTRACTIONS	Able to measure much higher flows than V-notch weir. Discharge equation more complicated than other types of weirs. Widely used for measuring high flow rates in channels suited to weirs.
	C. RECTANGULAR (SUPPRESSED) WEIR WITHOUT END CONTRACTIONS	Able to measure same range of flows as contracted rectangular weir, but easier to construct and has simpler discharge equation. However, width of weir crest must correspond to width of channel so use is restricted. May have problems obtaining adequate aeration of nappe.
	D. TRAPEZOIDAL (CIPOLLETTI) WEIR	Similar to rectangular contracted weir except that inclined ends result in simplified discharge equation. Less accurate than rectangular or V-notch weir, and therefore less often used.
	E. COMPOUND WEIR	Combination of any two types or sizes of above weirs to provide wide range of flows. Ambigious discharge curve in transition zone between two weirs.

(a)

PRIMARY MEASURING DEVICE		
GENERAL TYPE	**SPECIFIC TYPE**	**COMMENTS**
Advantages 1. Self-cleaning to a certain degree. 2. Relatively low head loss. 3. Accuracy less affected by approach velocity than weirs. **Disadvantages** 1. High cost. 2. Difficult to install.	A. PARSHALL FLUME	Most widely known and used flume for permanent installations. Available in throat widths ranging from 1 inch to 50 feet to cover most flows. Fairly difficult installation requiring a drop in the conduit invert.
	B. PALMER-BOWLUS FLUME	Flume designed to be easily installed in existing conduit. Good for portable or temporary installations, as no drop in conduit invert required. Widely used in sanitary field for measuring flows in manholes.
	C. H, HS, & HL FLUMES	Developed to measure agricultural runoff. Principle advantage is ability to measure wide range of flows with reasonable accuracy. Construction is fairly simple and flume is easily installed.
	D. CUTTHROAT FLUME	Similar to Parshall, except that flat bottom does not require drop in conduit invert. Can function well with high degrees of submergence. Flat bottom passes solids better than Parshall.
	E. TRAPEZOIDAL FLUME	Developed to measure flow in irrigation channels. Principle advantage is ability to measure wide range of flows and also maintain good accuracy at low flows.

(b)

SOURCE: Isco, Inc.

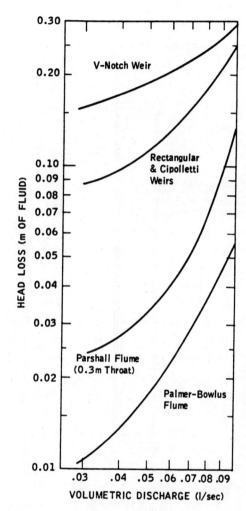

FIGURE 5-15 Relative head losses for water flows in different types of weirs and flumes. (*From Fluid Flow Pocket Handbook by Nicholas P. Cheremisinoff. Copyright 1984, Gulf Publishing Co., Houston, Texas. Used with permission. All rights reserved.*)

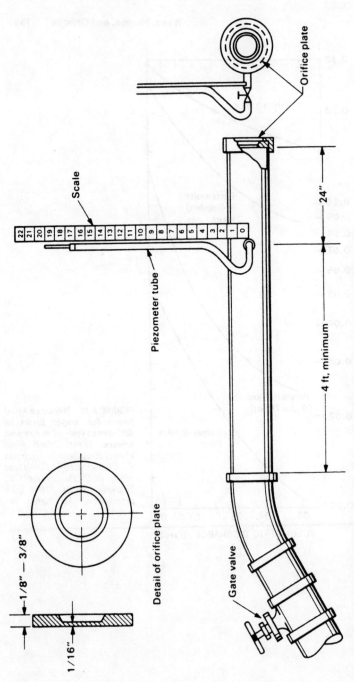

FIGURE 5-16 Essential details of the circular-orifice method used to measure pumping rates of a turbine pump. Discharge pipe must be level. (See Table 5-7.) (*Courtesy: Water Well Journal Publishing Co.*)

Orifice plate

Scale

Piezometer tube

24"

4 ft, minimum

Gate valve

1/8" — 3/8"

1/16"

Detail of orifice plate

TABLE 5-7   Flow Rates through Circular Orifice

Head of Water in Tube above Center of Orifice (inches)	4-inch Pipe, 2 1/2-inch Opening (gpm)	4-inch Pipe, 3-inch Opening (gpm)	6-inch Pipe, 3-inch Opening (gpm)	6-inch Pipe, 4-inch Opening (gpm)	8-inch Pipe, 4-inch Opening (gpm)	8-inch Pipe, 5-inch Opening (gpm)	8-inch Pipe, 6-inch Opening (gpm)	10-inch Pipe, 6-inch Opening (gpm)	10-inch Pipe, 7-inch Opening (gpm)	10-inch Pipe, 8-inch Opening (gpm)
5	55	89	–	–	–	–	–	–	–	–
6	60	97	82	158	144	240	390	–	–	–
7	65	105	88	171	156	260	420	370	540	830
8	69	112	94	182	166	275	450	395	580	880
9	73	119	100	193	176	295	475	420	610	940
10	77	126	106	204	186	310	500	440	640	990
12	85	138	115	223	205	340	550	480	700	1080
14	92	149	125	241	220	365	595	520	760	1170
16	98	159	132	258	235	390	635	555	810	1250
18	104	168	140	273	250	415	675	590	860	1330
20	110	178	150	288	265	440	710	620	910	1400
22	115	186	158	302	275	460	745	650	950	1470
25	122	198	168	322	295	490	795	690	1020	1560
30	134	217	182	353	325	540	870	760	1120	1710
35	145	235	198	380	355	580	940	820	1210	1850
40	155	251	210	405	370	620	1000	880	1290	1980
45	164	267	223	430	395	660	1060	930	1370	–
50	173	280	235	455	415	690	1120	980	1440	–
60	190	310	260	500	455	760	1230	1080	1580	–

SOURCE: *Water Well Journal* Publishing Co.

5-1   Venturi Meter, Nozzles, and Orifices

Venturi meter

The venturi meter is a common device for accurately measuring the discharge of pumps, particularly when a permanent meter installation is required. When the coefficient for the meter has been determined by actual calibration, and the meter is correctly installed and accurately read, the probable error in computing the discharge should be less than 1 percent.

As usually constructed, the meter consists of a converging portion, a throat having a diameter of approximately one-third the main pipe diameter, and a diverging portion to reduce loss of energy from turbulence. The length of the converging portion is usually 2 to 2½ times the diameter of the main pipe, while the best angle of divergence is about 10 degrees included angle.

For accurate results the distance from the nearest elbow or fitting to the entrance of the meter should be at least 10 times the diameter of the pipe. Otherwise, straightening vanes should be used to prevent spiral flow at entrance.

From a consideration of Bernoulli's theorem:

$$\text{Gallons per minute} = 3.118ca \sqrt{\frac{2gh}{R^4 - 1}}$$

where c = coefficient of discharge from calibration data. While
this coefficient may vary from about 0.94 to more
than unity, it is usually about 0.98.

a = area of entrance section where the upstream manometer connection is made, in square inches.

R = ratio of entrance to throat diameter = d/d_1.

g = acceleration of gravity (32.2 ft/sec^2).

h = $h_1 - h_2$ = difference in pressure between the entrance section and throat, as indicated by a manometer, in feet.

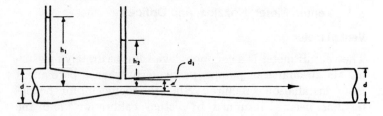

Venturi meter

In the illustration, the pressures h_1 and h_2 may be taken by
manometer as illustrated when the pressures are low. When
pressures are high, a differential mercury manometer, which
indicates the difference in pressure $h_1 - h_2$ directly, is most
often used. Gauges can also be used, but they can be read less
accurately than a manometer and do require calibration. In
commercial installations of venturi meters, instruments are
often installed that will continuously indicate, record, and/or
integrate the flow. They also require calibration, so, when con-
ducting a test, it is best to use a differential manometer connected directly to the meter to measure $h_1 - h_2$.

Nozzles

A nozzle is, in effect, the converging portion of a venturi tube. The water issues from the nozzle throat into the atmosphere. The pressure h_2, therefore, is atmospheric pressure. To calculate the flow from a nozzle, use the same formula as for the venturi meter. The head h in the formula will be the gauge reading h_1.

Orifices

Approximate discharge through orifice:

$$Q = 19.636\ Kd^2\sqrt{h}\ \sqrt{\cfrac{1}{1-\left(\cfrac{d}{D}\right)^4}} \qquad \text{where } d/D \text{ is greater than } .3$$

$$Q = 19.636\ Kd^2\ \sqrt{h} \qquad \text{where } d/D \text{ is less than } .3$$

where Q = flow, in gallons per minute
d = diameter of orifice or nozzle opening, in inches
h = head at orifice, in feet of liquid
D = diameter of pipe in which orifice is placed
K = discharge coefficient

This subsection courtesy of Fairbanks Morse Pump Corporation.

TABLE 5-8 Theoretical Discharge of Orifices, U.S. GPM

Head		Velocity of Discharge Feet per Second	Diameter of Orifice in Inches												
Lbs.	Feet		1/16	⅛	3/16	¼	⅜	½	⅝	¾	⅞	1	1⅛	1¼	1⅜
10	23.1	38.6	0.37	1.48	3.32	5.91	13.3	23.6	36.9	53.1	72.4	94.5	120	148	179
15	34.6	47.25	0.45	1.81	4.06	7.24	16.3	28.9	45.2	65.0	88.5	116.	147	181	219
20	46.2	54.55	0.52	2.09	4.69	8.35	18.8	33.4	52.2	75.1	102.	134.	169	209	253
25	57.7	61.0	0.58	2.34	5.25	9.34	21.0	37.3	58.3	84.0	114.	149.	189	234	283
30	69.3	66.85	0.64	2.56	5.75	10.2	23.0	40.9	63.9	92.0	125.	164.	207	256	309
35	80.8	72.2	0.69	2.77	6.21	11.1	24.8	44.2	69.0	99.5	135.	177.	224	277	334
40	92.4	77.2	0.74	2.96	6.64	11.8	26.6	47.3	73.8	106.	145.	189.	239	296	357
45	103.9	81.8	0.78	3.13	7.03	12.5	28.2	50.1	78.2	113.	153.	200.	253	313	379
50	115.5	86.25	0.83	3.30	7.41	13.2	29.7	52.8	82.5	119.	162.	211.	267	330	399
55	127.0	90.4	0.87	3.46	7.77	13.8	31.1	55.3	86.4	125.	169.	221.	280	346	418
60	138.6	94.5	0.90	3.62	8.12	14.5	32.5	57.8	90.4	130.	177.	231.	293	362	438
65	150.1	98.3	0.94	3.77	8.45	15.1	33.8	60.2	94.0	136.	184.	241.	305	376	455
70	161.7	102.1	0.98	3.91	8.78	15.7	35.2	62.5	97.7	141.	191.	250.	317	391	473
75	173.2	105.7	1.01	4.05	9.08	16.2	36.4	64.7	101.	146.	198.	259.	327	404	489
80	184.8	109.1	1.05	4.18	9.39	16.7	37.6	66.8	104.	150.	205.	267.	338	418	505
85	196.3	112.5	1.08	4.31	9.67	17.3	38.8	68.9	108.	155.	211.	276.	349	431	521
90	207.9	115.8	1.11	4.43	9.95	17.7	39.9	70.8	111.	160.	217.	284.	359	443	536
95	219.4	119.0	1.14	4.56	10.2	18.2	41.0	72.8	114.	164.	223.	292.	369	456	551
100	230.9	122.0	1.17	4.67	10.5	18.7	42.1	74.7	117.	168.	229.	299.	378	467	565
105	242.4	125.0	1.20	4.79	10.8	19.2	43.1	76.5	120.	172.	234.	306.	388	479	579
110	254.0	128.0	1.23	4.90	11.0	19.6	44.1	78.4	122.	176.	240.	314.	397	490	593
115	265.5	130.9	1.25	5.01	11.2	20.0	45.1	80.1	125.	180.	245.	320.	406	501	606
120	277.1	133.7	1.28	5.12	11.5	20.5	46.0	81.8	128.	184.	251.	327.	414	512	619
125	288.6	136.4	1.31	5.22	11.7	20.9	47.0	83.5	130.	188.	256.	334.	423	522	632
130	300.2	139.1	1.33	5.33	12.0	21.3	48.0	85.2	133.	192.	261.	341.	432	533	645
135	311.7	141.8	1.36	5.43	12.2	21.7	48.9	86.7	136.	195.	266.	347.	439	543	656
140	323.3	144.3	1.38	5.53	12.4	22.1	49.8	88.4	138.	199.	271.	354.	448	553	668
145	334.8	146.9	1.41	5.62	12.6	22.5	50.6	89.9	140.	202.	275.	360.	455	562	680
150	346.4	149.5	1.43	5.72	12.9	22.9	51.5	91.5	143.	206.	280.	366.	463	572	692
175	404.1	161.4	1.55	6.18	13.9	24.7	55.6	98.8	154.	222.	302.	395.	500	618	747
200	461.9	172.6	1.65	6.61	14.8	26.4	59.5	106.	165.	238.	323.	423.	535	660	799
250	577.4	193.0	1.85	7.39	16.6	29.6	66.5	118.	185.	266.	362.	473.	598	739	894
300	692.8	211.2	2.02	8.08	18.2	32.4	72.8	129.	202.	291.	396.	517.	655	808	977

NOTE—To determine actual discharge multiply values from table by coefficient of discharge.

TABLE 5-8 Theoretical Discharge of Orifices, U.S. GPM (Continued)

Head		Velocity of Discharge Feet Per Second	Diameter of Orifice in Inches												
Lbs.	Feet		1½	1¾	2	2¼	2½	2¾	3	3½	4	4½	5	5½	6
10	23.1	38.6	213	289	378	479	591	714	851	1158	1510	1915	2365	2855	3405
15	34.6	47.25	260	354	463	585	723	874	1041	1418	1850	2345	2890	3490	4165
20	46.2	54.55	301	409	535	676	835	1009	1203	1638	2135	2710	3340	4040	4819
25	57.7	61.0	336	458	598	756	934	1128	1345	1830	2385	3025	3730	4510	5380
30	69.3	66.85	368	501	655	828	1023	1236	1473	2005	2615	3315	4090	4940	5895
35	80.8	72.2	398	541	708	895	1106	1335	1591	2168	2825	3580	4415	5340	6370
40	92.4	77.2	425	578	756	957	1182	1428	1701	2315	3020	3830	4725	5710	6810
45	103.9	81.8	451	613	801	1015	1252	1512	1802	2455	3200	4055	5000	6050	7210
50	115.5	86.25	475	647	845	1070	1320	1595	1900	2590	3375	4275	5280	6380	7600
55	127.0	90.4	498	678	886	1121	1385	1671	1991	2710	3540	4480	5530	6690	7970
60	138.6	94.5	521	708	926	1172	1447	1748	2085	2835	3700	4685	5790	6980	8330
65	150.1	98.3	542	737	964	1220	1506	1819	2165	2950	3850	4875	6020	7270	8670
70	161.7	102.1	563	765	1001	1267	1565	1888	2250	3065	4000	5060	6250	7560	9000
75	173.2	105.7	582	792	1037	1310	1619	1955	2330	3170	4135	5240	6475	7820	9320
80	184.8	109.1	602	818	1070	1354	1672	2020	2405	3280	4270	5410	6690	8080	9630
85	196.3	112.5	620	844	1103	1395	1723	2080	2480	3375	4400	5575	6890	8320	9920
90	207.9	115.8	638	868	1136	1436	1773	2140	2550	3475	4530	5740	7090	8560	10210
95	219.4	119.0	655	892	1168	1476	1824	2200	2625	3570	4655	5900	7290	8800	10500
100	230.9	122.0	672	915	1196	1512	1870	2255	2690	3660	4775	6050	7470	9030	10770
105	242.4	125.0	689	937	1226	1550	1916	2312	2755	3750	4890	6200	7650	9250	11020
110	254.0	128.0	705	960	1255	1588	1961	2366	2820	3840	5010	6350	7840	9470	11300
115	265.5	130.9	720	980	1282	1621	2005	2420	2885	3930	5120	6490	8010	9680	11550
120	277.1	133.7	736	1002	1310	1659	2050	2470	2945	4015	5225	6630	8180	9900	11800
125	288.6	136.4	751	1022	1338	1690	2090	2520	3005	4090	5340	6760	8350	10100	12030
130	300.2	139.1	767	1043	1365	1726	2132	2575	3070	4175	5450	6900	8530	10300	12290
135	311.7	141.8	780	1063	1390	1759	2173	2620	3125	4250	5550	7030	8680	10400	12510
140	323.3	144.3	795	1082	1415	1790	2212	2670	3180	4330	5650	7160	8850	10600	12730
145	334.8	146.9	809	1100	1440	1820	2250	2715	3235	4410	5740	7280	8990	10880	12960
150	346.4	149.5	824	1120	1466	1853	2290	2760	3295	4485	5850	7410	9150	11070	13200
175	404.1	161.4	890	1210	1582	2000	2473	2985	3560	4840	6310	8000	9890	11940	14250
200	461.9	172.6	950	1294	1691	2140	2645	3190	3800	5175	6750	8550	10580	12770	15220
250	577.4	193.0	1063	1447	1891	2392	2955	3570	4250	5795	7550	9570	11820	14290	17020
300	692.8	211.2	1163	1582	2070	2615	3235	3900	4650	6330	8260	10480	12940	15620	18610

NOTE—To determine actual discharge multiply values from table by coefficient of discharge.

SOURCE: Fairbanks Morse Pump Corporation.

TABLE 5-9   Open-Flow Nozzles—Dimensions and Approximate Capacities

NOZZLE DIA. IN.	NOZZLE LENGTH, IN.		APPROXIMATE MAXIMUM CAPACITY			
			PARABOLIC		KENNISON	
	PARABOLIC	KENNISON	MGD	CFS	MGD	CFS
6	28	12	0.27	0.42	0.27	0.42
8	35	16	0.57	0.88	0.45	0.70
10	43	20	0.97	1.50	0.84	1.31
12	50	24	1.50	2.32	1.25	1.94
16	66	32	2.92	4.52	2.71	4.19
20	81	40	4.91	7.60	4.51	6.97
24	96	48	7.47	11.6	7.46	11.5
30	119	60	12.5	19.4	11.6	17.9
36	142	72	19.4	30.1	19.4	30.1

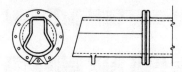

A. Kennison Nozzle ($Q \alpha H$)

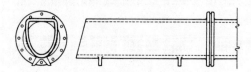

B. Parabolic Nozzle ($Q \alpha H^2$)

SOURCE: Isco, Inc.

TABLE 5-10   Channel Section Geometric Elements

Channel type	Area A	Wetted perimeter P	Hydraulic radius R	Top width T	Hydraulic depth D	Section factor Z
(a)	by	$b + 2y$	$\dfrac{by}{b+2y}$	b	y	$by^{1.5}$
(b)	$(b+zy)y$	$b + 2y\sqrt{1+z^2}$	$\dfrac{(b+zy)y}{b+2y\sqrt{1+z^2}}$	$b + 2zy$	$\dfrac{(b+zy)y}{b+2zy}$	$\dfrac{[(b+zy)y]^{1.5}}{\sqrt{b+2zy}}$
(c)	zy^2	$2y\sqrt{1+z^2}$	$\dfrac{zy}{2\sqrt{1+z^2}}$	$2zy$	$\tfrac{1}{2}y$	$\dfrac{\sqrt{2}}{2}zy^{2.5}$
(d)	$\tfrac{2}{3}Ty$	$T + \dfrac{8}{3}\dfrac{y^{2\,*}}{T}$	$\dfrac{2T^2y}{3T^2+8y^2}$	$\dfrac{3}{2}\dfrac{A}{y}$	$\tfrac{2}{3}y$	$\tfrac{2}{3}\sqrt{6}\,Ty^{1.5}$
(e)	$\tfrac{1}{8}(\theta - \sin\theta)\,d_0^2$	$\tfrac{1}{2}\theta\, d_0$	$\dfrac{1}{4}\left(1 - \dfrac{\sin\theta}{\theta}\right)d_0$	$2\sqrt{y(d_0 - y)}$	$\dfrac{1}{8}\left(\dfrac{\theta - \sin\theta}{\sin\frac{1}{2}\theta}\right)d_0$	$\dfrac{\sqrt{2}(\theta - \sin\theta)^{1.5}}{32\sqrt{\sin\frac{1}{2}\theta}}\,d_0^{2.5}$

*Satisfactory approximation when $0 < 4y/T \le 1$.

For $4y/T > 1$ $P = \dfrac{T}{2}\left[\sqrt{1+\left(\dfrac{4y}{T}\right)^2} + \dfrac{T}{4y}\ln\left(\dfrac{4y}{T} + \sqrt{1+\left(\dfrac{4y}{T}\right)^2}\,\right)\right]$

SOURCE: *Open-Channel Hydraulics* by R. H. French. Copyright 1985, McGraw-Hill, Inc.

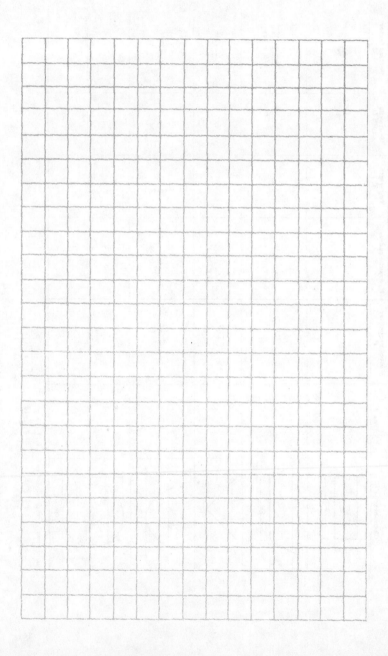

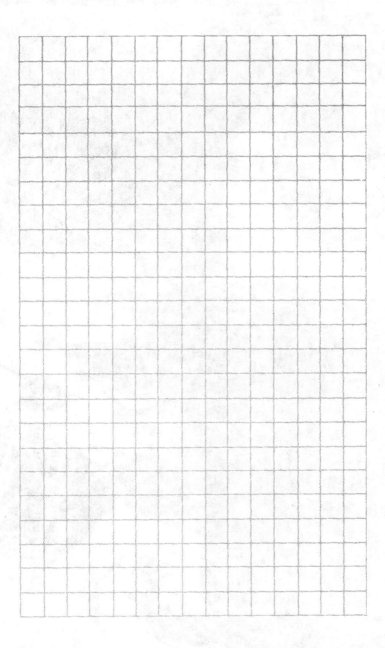

6

Flow in Pipes

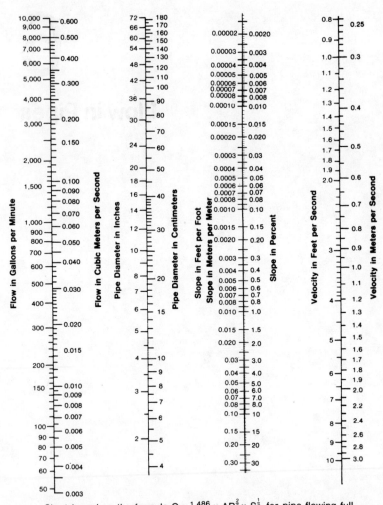

Chart based on the formula $Q = \frac{1.486}{n} \times AR^{\frac{2}{3}} \times S^{\frac{1}{2}}$ for pipe flowing full.

FIGURE 6-1 Manning formula pipe flow chart (English/metric units). $n =$ 0.009. *[Courtesy: Water & Sewer Works, Sept. 1977 (Now — Water Engineering & Management).]*

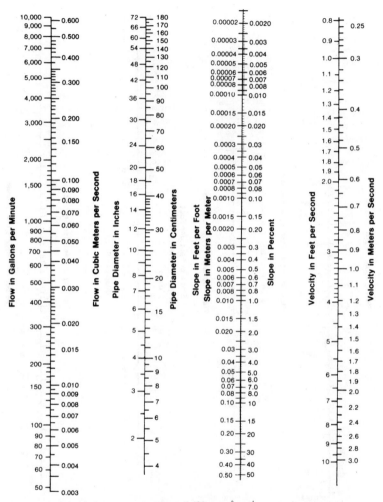

Chart based on the formula $Q = \frac{1.486}{n} \times AR^{\frac{2}{3}} \times S^{\frac{1}{2}}$ for pipe flowing full.

FIGURE 6-2 Manning formula pipe flow chart (English/metric units). $n = 0.010$. *[Courtesy: Water & Sewer Works, Sept. 1977 (Now — Water Engineering & Management).]*

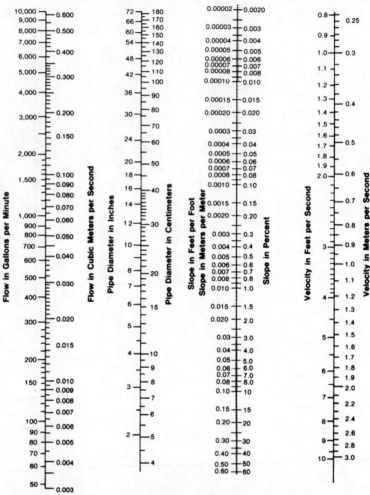

Chart based on the formula $Q = \frac{1.486}{n} \times AR^{\frac{2}{3}} \times S^{\frac{1}{2}}$ for pipe flowing full.

FIGURE 6-3 Manning formula pipe flow chart (English/metric units). $n =$ 0.011. *[Courtesy: Water & Sewer Works, Sept. 1977 (Now — Water Engineering & Management).]*

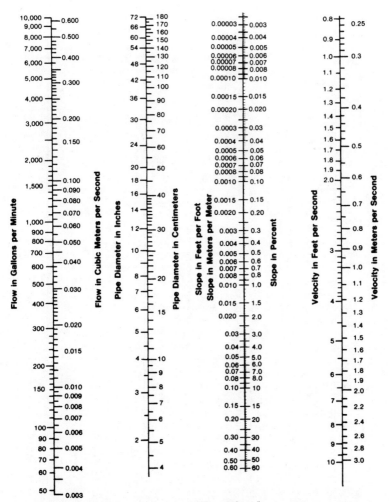

Chart based on the formula $Q = \frac{1.486}{n} \times AR^{\frac{2}{3}} \times S^{\frac{1}{2}}$ for pipe flowing full.

FIGURE 6-4 Manning formula pipe flow chart (English/metric units). $n = 0.012$. *[Courtesy: Water & Sewer Works, Sept. 1977 (Now — Water Engineering & Management).]*

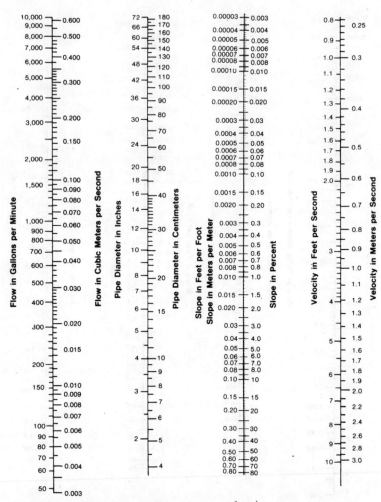

Chart based on the formula $Q = \frac{1.486}{n} \times AR^{\frac{2}{3}} \times S^{\frac{1}{2}}$ for pipe flowing full.

FIGURE 6-5   Manning formula pipe flow chart (English/metric units). $n =$ 0.013. *[Courtesy: Water & Sewer Works, Sept. 1977 (Now — Water Engineering & Management).]*

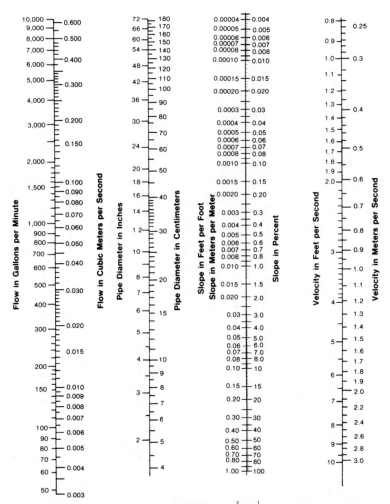

Chart based on the formula $Q = \frac{1.486}{n} \times AR^{\frac{2}{3}} \times S^{\frac{1}{2}}$ for pipe flowing full.

FIGURE 6-6 Manning formula pipe flow chart (English/metric units). $n = 0.015$. *[Courtesy: Water & Sewer Works, Sept. 1977 (Now — Water Engineering & Management).]*

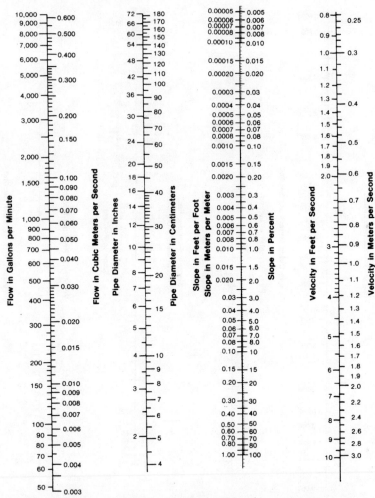

Chart based on the formula $Q = \frac{1.486}{n} \times AR^{\frac{2}{3}} \times S^{\frac{1}{2}}$ for pipe flowing full.

FIGURE 6-7 Manning formula pipe flow chart (English/metric units). $n =$ 0.017. *[Courtesy: Water & Sewer Works, Sept. 1977 (Now — Water Engineering & Management).]*

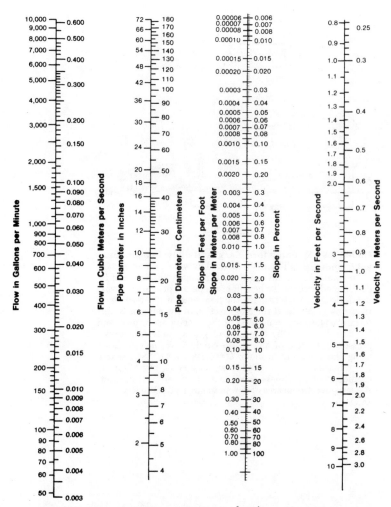

Chart based on the formula $Q = \frac{1.486}{n} \times AR^{\frac{2}{3}} \times S^{\frac{1}{2}}$ for pipe flowing full.

FIGURE 6-8 Manning formula pipe flow chart (English/metric units). $n = 0.019$. *[Courtesy: Water & Sewer Works, Sept. 1977 (Now — Water Engineering & Management).]*

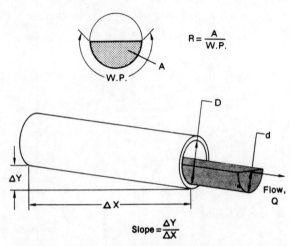

$$R = \frac{A}{W.P.}$$

$$Slope = \frac{\Delta Y}{\Delta X}$$

FIGURE 6-9 Manning formula: gravity flow in open channel (round pipe). There is a method by which, under certain circumstances, the rate of flow in an open channel can be determined without the benefit of a separate primary measuring device. In this technique the flow conduit itself serves as the primary device. If the cross section of the conduit is uniform, the slope and roughness of the conduit are known, and the flow is moved by the force of gravity only (not under pressure), the rate of flow in the conduit may be calculated using the Manning formula: $Q = (1.49/n)\ AR^{2/3}S^{1/2}$, where Q = quantity of flow in cubic feet per second, n = Manning coefficient of roughness dependent on material of conduit, A = cross-sectional area of flow in square feet. *(Courtesy: Isco, Inc.)*

TABLE 6-1 Values of the Manning Roughness Coefficient *n*

DESCRIPTION OF CHANNEL	MINIMUM	NORMAL	MAXIMUM
I. Closed conduit - Partly full			
A. Metal			
1. Steel			
a. Lockbar and welded	0.010	0.012	0.014
b. Riveted and spiral	0.013	0.016	0.017
2. Cast Iron			
a. Coated	0.010	0.013	0.014
b. Uncoated	0.011	0.014	0.016
3. Wrought Iron			
a. Black	0.012	0.014	0.015
b. Galvanized	0.013	0.016	0.017
4. Corrugated			
a. Subdrain	0.017	0.019	0.021
b. Storm drain	0.021	0.024	0.030
B. Nonmetal			
1. Acrylic	0.008	0.009	0.010
2. Glass	0.009	0.010	0.013
3. Wood			
a. Stave	0.010	0.012	0.014
b. Laminated, treated	0.015	0.017	0.020
4. Clay			
a. Common drainage tile	0.011	0.013	0.017
b. Vitrified sewer	0.011	0.014	0.017
c. Vitrified sewer with manholes, inlets, etc.	0.013	0.015	0.017
5. Brick			
a. Glazed	0.011	0.013	0.015
b. Lined with cement	0.012	0.015	0.017
6. Concrete			
a. Culvert, straight and free of debris	0.010	0.011	0.013
b. Culvert with bends, connections, and some debris	0.011	0.013	0.014
c. Sewer with manholes, inlet, etc., straight	0.013	0.015	0.017
d. Unfinished, steel form	0.012	0.013	0.014
e. Unfinished, smooth wood form	0.012	0.014	0.016
f. Unfinished, rough wood form	0.015	0.017	0.020
7. Sanitary sewers coated with sewage slimes	0.012	0.013	0.016

TABLE 6-1 Values of the Manning Roughness Coefficient *n (Continued)*

DESCRIPTION OF CHANNEL	MINIMUM	NORMAL	MAXIMUM
8. Paved invert, sewer, smooth bottom	0.016	0.019	0.020
9. Rubble masonry, cemented	0.018	0.025	0.030
II. Lined or Built-up Channels			
A. Metal			
1. Smooth steel surface			
a. Painted	0.011	0.012	0.014
b. Unpainted	0.012	0.013	0.017
2. Corrugated	0.021	0.025	0.030
B. Nonmetal			
1. Cement			
a. Neat surface	0.010	0.011	0.013
b. Mortar	0.011	0.013	0.015
2. Concrete			
a. Trowel finish	0.011	0.013	0.015
b. Float finish	0.013	0.015	0.016
c. Finished, with gravel on bottom	0.015	0.017	0.020
d. Unfinished	0.014	0.017	0.020
3. Wood			
a. Planed, untreated	0.010	0.012	0.014
b. Planed, creosoted	0.011	0.012	0.015
c. Unplaned	0.011	0.013	0.015
d. Plank with battens	0.012	0.015	0.018
4. Brick			
a. Glazed	0.011	0.013	0.015
b. In cement mortar	0.012	0.015	0.018
5. Masonry			
a. Cemented rubble	0.017	0.025	0.030
b. Dry rubble	0.023	0.032	0.035
6. Asphalt			
a. Smooth	0.013	0.013	——
b. Rough	0.016	0.016	——
7. Vegetal lining	0.030	——	0.500
III. Excavated or Dredged			
A. Earth, straight and uniform	0.016	0.022	0.035
B. Earth, winding and sluggish	0.023	0.030	0.040
C. Rock cuts	0.030	0.040	0.050
D. Unmaintained channels	0.040	0.070	0.140

TABLE 6-1 Values of the Manning Roughness Coefficient *n (Continued)*

DESCRIPTION OF CHANNEL	MINIMUM	NORMAL	MAXIMUM

IV. Natural Channels
    (Minor streams, top width
    at flood 100 ft.)
  A. Fairly regular section
	0.030	0.050	0.070

B. Irregular section with pools	0.040	0.070	0.100

SOURCE: Isco, Inc.

TABLE 6-2 Area of Flow and Hydraulic Radius for Various Flow Depths

d / D	A / D²	R / D	d / D	A / D²	R / D	d / D	A / D²	R / D
0.01	0.0013	0.0066	0.36	0.2546	0.1978	0.71	0.5964	0.2973
0.02	0.0037	0.0132	0.37	0.2642	0.2020	0.72	0.6054	0.2984
0.03	0.0069	0.0197	0.38	0.2739	0.2061	0.73	0.6143	0.2995
0.04	0.0105	0.0262	0.39	0.2836	0.2102	0.74	0.6231	0.3006
0.05	0.0147	0.0326	0.40	0.2934	0.2142	0.75	0.6318	0.3017
0.06	0.0192	0.0389	0.41	0.3032	0.2181	0.76	0.6404	0.3025
0.07	0.0242	0.0451	0.42	0.3130	0.2220	0.77	0.6489	0.3032
0.08	0.0294	0.0513	0.43	0.3229	0.2257	0.78	0.6573	0.3037
0.09	0.0350	0.0574	0.44	0.3328	0.2294	0.79	0.6655	0.3040
0.10	0.0409	0.0635	0.45	0.3428	0.2331	0.80	0.6736	0.3042
0.11	0.0470	0.0695	0.46	0.3527	0.2366	0.81	0.6815	0.3044
0.12	0.0534	0.0754	0.47	0.3627	0.2400	0.82	0.6893	0.3043
0.13	0.0600	0.0813	0.48	0.3727	0.2434	0.83	0.6969	0.3041
0.14	0.0668	0.0871	0.49	0.3827	0.2467	0.84	0.7043	0.3038
0.15	0.0739	0.0929	0.50	0.3927	0.2500	0.85	0.7115	0.3033
0.16	0.0811	0.0986	0.51	0.4027	0.2531	0.86	0.7186	0.3026
0.17	0.0885	0.1042	0.52	0.4127	0.2561	0.87	0.7254	0.3017
0.18	0.0961	0.1097	0.53	0.4227	0.2591	0.88	0.7320	0.3008
0.19	0.1039	0.1152	0.54	0.4327	0.2620	0.89	0.7384	0.2996
0.20	0.1118	0.1206	0.55	0.4426	0.2649	0.90	0.7445	0.2980
0.21	0.1199	0.1259	0.56	0.4526	0.2676	0.91	0.7504	0.2963
0.22	0.1281	0.1312	0.57	0.4625	0.2703	0.92	0.7560	0.2944
0.23	0.1365	0.1364	0.58	0.4723	0.2728	0.93	0.7612	0.2922
0.24	0.1449	0.1416	0.59	0.4822	0.2753	0.94	0.7662	0.2896
0.25	0.1535	0.1466	0.60	0.4920	0.2776	0.95	0.7707	0.2864
0.26	0.1623	0.1516	0.61	0.5018	0.2797	0.96	0.7749	0.2830
0.27	0.1711	0.1566	0.62	0.5115	0.2818	0.97	0.7785	0.2787
0.28	0.1800	0.1614	0.63	0.5212	0.2839	0.98	0.7816	0.2735
0.29	0.1890	0.1662	0.64	0.5308	0.2860	0.99	0.7841	0.2665
0.30	0.1982	0.1709	0.65	0.5404	0.2881	1.00	0.7854	0.2500
0.31	0.2074	0.1755	0.66	0.5499	0.2899			
0.32	0.2167	0.1801	0.67	0.5594	0.2917			
0.33	0.2260	0.1848	0.68	0.5687	0.2935			
0.34	0.2355	0.1891	0.69	0.5780	0.2950			
0.35	0.2450	0.1935	0.70	0.5872	0.2962			

A = Area of flow
R = Hydraulic radius
SOURCE: Isco, Inc.

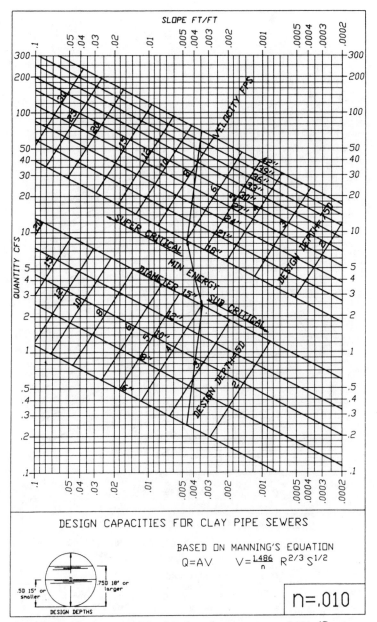

DESIGN CAPACITIES FOR CLAY PIPE SEWERS

BASED ON MANNING'S EQUATION

$$Q = AV \qquad V = \frac{1.486}{n} R^{2/3} S^{1/2}$$

$$n = .010$$

FIGURE 6-10 **Design capacities for clay pipe sewers.** $n = 0.010$. *(Courtesy: National Clay Pipe Institute.)*

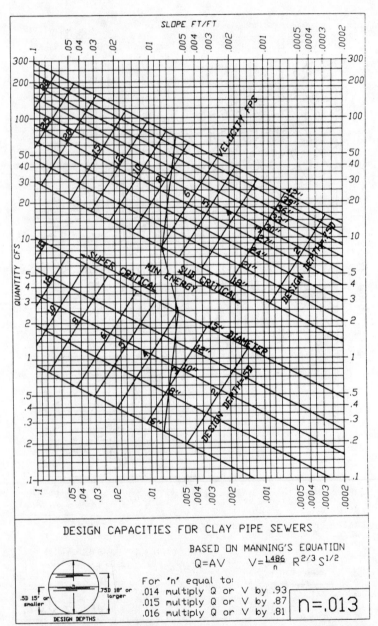

FIGURE 6-11 Design capacities for clay pipe sewers. $n = 0.013$. *(Courtesy: National Clay Pipe Institute.)*

TABLE 6-3 Hydraulic Properties of Clay Pipe at Design Depth

DIA. (in.)	DIA. (ft.)	DESIGN DEPTH	A	R	R2/3	AR2/3	%CAP. INCR.*
6	.5	.5D	.10	.125	.250	.025	
8	.667	.5D	.17	.167	.303	.052	108
10	.833	.5D	.27	.208	.350	.094	81
12	1.0	.5D	.39	.250	.397	.155	65
15	1.25	.5D	.61	.312	.460	.281	81
18	1.5	.75D	1.42	.453	.590	.838	198
21	1.75	.75D	1.94	.528	.654	1.27	52
24	2.0	.75D	2.53	.604	.714	1.81	43
27	2.25	.75D	3.20	.680	.773	2.47	36
30	2.5	75D	3.95	.755	.829	3.27	32
33	2.75	.75D	4.78	.830	.883	4.22	29
36	3.0	.75D	5.69	.906	.936	5.33	26
39	3.25	.75D	6.68	.982	.988	6.60	24
42	3 5	.75D	7.74	1.057	1.018	8.03	22

*Percent capacity increase above the capacity of the previous smaller diameter.

SOURCE: National Clay Pipe Institute

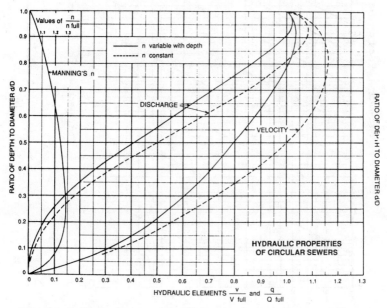

FIGURE 6-12 **Hydraulic properties of circular sewers.** *(Courtesy: National Clay Pipe Institute.)*

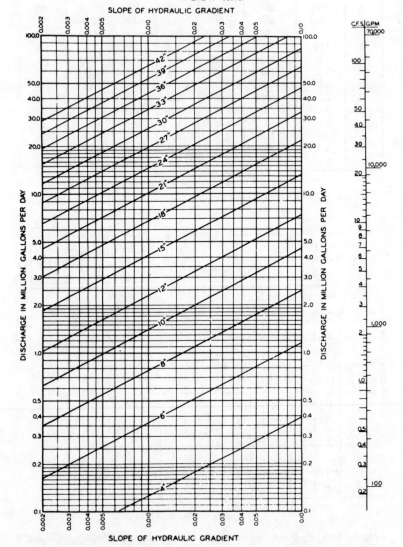

FIGURE 6-13 Discharge of circular pipes — flowing full. *(Courtesy: National Clay Pipe Institute.)*

TABLE 6-4 Hydraulic Properties of Clay Pipe

Diameter (inches)	Diameter (feet)	Area (square feet)	Hydraulic Radius (flowing full or half full) (feet)
4	0.333	0.087	0.083
6	0.500	0.196	0.125
8	0.667	0.349	0.167
10	0.833	0.545	0.208
12	1.000	0.785	0.250
15	1.250	1.227	0.313
18	1.500	1.767	0.375
21	1.750	2.405	0.437
24	2.000	3.142	0.500
27	2.250	3.976	0.563
30	2.500	4.909	0.625
33	2.750	5.940	0.688
36	3.000	7.068	0.750
39	3.250	8.300	0.813
42	3.500	9.621	0.875

SOURCE: National Clay Pipe Institute.

TABLE 6-5 Relative Carrying Capacities of Clay Pipe at Any Given Slope

12-inch pipe used as 100 percent.

Pipe Size (inches)	Relative Carrying Capacity in Per Cent	Pipe Size (inches)	Relative Carrying Capacity in Per Cent
4	5.3	21	445
6	15.7	24	635
8	34.0	27	870
10	61.4	30	1152
12	100	33	1484
15	181	36	1872
18	295	39	2319
		42	2820

SOURCE: National Clay Pipe Institute.

TABLE 6-6   Velocity and Discharge in Sewers and Drainage Pipes (Based on Kutter's Formula, Pipes Flowing Full)

Diam. of Pipe in In.	N — .011		N — .013		Diam. of Pipe in In.	N — .011		N — .013	
	Velocity in feet per second	Discharge gals. per minute	Velocity in feet per second	Discharge gals. per minute		Velocity in feet per second	Discharge gals. per minute	Velocity in feet per second	Discharge gals. per minute
SLOPE — 0.2 FEET PER HUNDRED FEET					SLOPE — 0.3 FEET PER HUNDRED FEET				
6	1.40	124	1.12	99	6	1.68	148	1.34	118
8	1.76	274	1.41	219	8	2.07	322	1.69	263
10	2.07	506	1.65	403	10	2.48	606	1.99	486
12	2.37	835	1.94	684	12	2.84	1,001	2.32	818
15	2.80	1,545	2.28	1,259	15	3.34	1,844	2.71	1,496
18	3.20	2,545	2.60	2,067	18	3.82	3,038	3.10	2,465
21	3.57	3,859	2.92	3,156	21	4.29	4,637	3.50	3,783
24	3.90	5,513	3.20	4,524	24	4.68	6,616	3.83	5,415
27	4.27	7,639	3.51	6,279	27	5.10	9,124	4.20	7,514
30	4.58	10,117	3.87	8,549	30	5.48	12,105	4.59	10,139
33	4.90	13,093	4.02	10,741	33	5.86	15,658	4.82	12,879
36	5.18	16,476	4.28	13,614	36	6.20	19,721	5.13	16,318

Diam. of Pipe in In.	N — .011		N — .013		Diam. of Pipe in In.	N — .011		N — .013	
	Velocity in feet per second	Discharge gals. per minute	Velocity in feet per second	Discharge gals. per minute		Velocity in feet per second	Discharge gals. per minute	Velocity in feet per second	Discharge gals. per minute
SLOPE — 0.4 FEET PER HUNDRED FEET					SLOPE — 0.5 FEET PER HUNDRED FEET				
6	1.96	173	1.56	138	6	2.24	198	1.78	157
8	2.38	370	1.97	306	8	2.68	417	2.24	351
10	2.88	705	2.33	570	10	3.28	803	2.67	654
12	3.31	1,167	2.70	952	12	3.78	1,332	3.07	1,082
15	3.87	2,136	3.14	1,733	15	4.40	2,428	3.57	1,971
18	4.44	3,531	3.60	2,863	18	5.05	4,015	4.10	3,260
21	5.01	5,416	4.08	4,410	21	5.72	6,186	4.66	5,040
24	5.46	7,719	4.45	6,291	24	6.23	8,807	5.07	7,168
27	5.93	10,609	4.88	8,730	27	6.76	12,094	5.56	9,947
30	6.38	14,093	5.30	11,708	30	7.28	16,081	6.01	13,276
33	6.82	18,223	5.61	14,990	33	7.77	20,769	6.40	17,107
36	7.21	22,934	5.97	18,989	36	8.22	26,146	6.81	21,661

TABLE 6-6 Velocity and Discharge in Sewers and Drainage Pipes (Based on Kutter's Formula, Pipes Flowing Full) *(Continued)*

Diam. of Pipe in In.	N — .011 Velocity in feet per second	N — .011 Discharge gals. per minute	N — .013 Velocity in feet per second	N — .013 Discharge gals. per minute

SLOPE — 0.6 FEET PER HUNDRED FEET

6	2.42	214	1.92	169
8	2.92	453	2.42	380
10	3.56	871	2.89	707
12	4.10	1,445	3.32	1,171
15	4.79	2,644	3.90	2,153
18	5.48	4,358	4.45	3,539
21	6.18	6,684	5.04	5,451
24	6.73	9,514	5.48	7,747
27	7.32	13,096	6.02	10,770
30	7.89	17,428	6.51	14,380
33	8.43	22,533	6.93	18,524
36	8.92	28,373	7.36	23,411

SLOPE — 0.7 FEET PER HUNDRED FEET

Diam.	N—.011 Vel	Disch	N—.013 Vel	Disch
6	2.61	231	2.07	182
8	3.15	489	2.61	409
10	3.84	940	3.11	761
12	4.42	1,558	3.58	1,260
15	5.18	2,859	4.22	2,330
18	5.91	4,700	4.80	3,817
21	6.64	7,181	5.42	5,862
24	7.23	10,221	5.89	8,327
27	7.88	14,098	6.48	11,593
30	8.50	18,776	7.00	15,462
33	9.09	24,297	7.46	19,940
36	9.62	30,599	7.91	25,160

SLOPE — 0.8 FEET PER HUNDRED FEET

Diam.	N—.011 Vel	Disch	N—.013 Vel	Disch
6	2.79	246	2.21	195
8	3.38	525	2.79	439
10	4.12	1,008	3.33	814
12	4.74	1,671	3.83	1,349
15	5.56	3,069	4.54	2,506
18	6.34	5,042	5.15	4,095
21	7.10	7,678	5.79	6,262
24	7.73	10,928	6.30	8,906
27	8.44	15,100	6.94	12,416
30	9.11	20,123	7.49	16,545
33	9.75	26,061	7.99	21,357
36	10.31	32,794	8.46	26,910

SLOPE — 0.9 FEET PER HUNDRED FEET

Diam.	N—.011 Vel	Disch	N—.013 Vel	Disch
6	2.98	263	2.36	208
8	3.61	561	2.98	468
10	4.39	1,074	3.55	867
12	5.06	1,784	4.08	1,438
15	5.94	3,279	4.86	2,683
18	6.77	5,384	5.50	4,374
21	7.56	8,176	6.16	6,662
24	8.22	11,621	6.71	9,486
27	9.00	16,101	7.40	13,239
30	9.72	21,471	7.98	17,627
33	10.40	27,799	8.52	22,774
36	11.00	34,989	9.01	28,659

TABLE 6-6 Velocity and Discharge in Sewers and Drainage Pipes (Based on Kutter's Formula, Pipes Flowing Full) *(Continued)*

	N — .011		N — .013	
Diam. of Pipe in In.	Velocity in feet per second	Discharge gals. per minute	Velocity in feet per second	Discharge gals. per minute
SLOPE — 1.0 FEET PER HUNDRED FEET				
6	3.17	280	2.51	221
8	3.84	597	3.17	497
10	4.66	1,140	3.76	920
12	5.38	1,896	4.33	1,526
15	6.32	3,489	5.18	2,860
18	7.20	5,726	5.85	4,652
21	8.02	8,673	6.53	7,062
24	8.71	12,314	7.12	10,066
27	9.56	17,103	7.85	14,044
30	10.33	22,818	8.47	18,710
33	11.05	29,536	9.05	24,190
36	11.69	37,184	9.56	30,408

	N — .011		N — .013	
Diam. of Pipe in In.	Velocity in feet per second	Discharge gals. per minute	Velocity in feet per second	Discharge gals. per minute
SLOPE — 2.0 FEET PER HUNDRED FEET				
6	4.48	396	3.56	314
8	5.57	866	4.48	702
10	6.60	1,616	5.35	1,310
12	7.53	2,654	6.13	2,161
15	8.96	4,946	7.26	4,008
18	10.20	8,111	8.29	6,592
21	11.48	12,415	9.35	10,112
24	12.43	17,572	10.17	14,378
27	13.59	24,313	11.09	19,840
30	14.50	32,029	11.89	26,264
33	15.71	41,993	12.87	34,401
36	16.45	52,356	13.55	43,100

	N — .011		N — .013	
Diam. of Pipe in In.	Velocity in feet per second	Discharge gals. per minute	Velocity in feet per second	Discharge gals. per minute
SLOPE — 3.0 FEET PER HUNDRED FEET				
6	5.49	485	4.35	383
8	6.72	1,045	5.49	860
10	8.14	1,992	6.53	1,598
12	9.35	3,296	7.51	2,647
15	10.94	5,984	8.40	4,637
18	12.58	10,002	10.15	8,071
21	14.04	15,184	11.44	12,372
24	15.17	21,446	12.40	17,530
27	16.66	29,806	13.68	24,474
30	17.90	39,540	14.67	32,405
33	19.17	51,241	15.80	42,233
36	20.23	64,347	16.66	52,992

	N — .011		N — .013	
Diam. of Pipe in In.	Velocity in feet per second	Discharge gals. per minute	Velocity in feet per second	Discharge gals. per minute
SLOPE — 4.0 FEET PER HUNDRED FEET				
6	6.34	560	5.05	445
8	7.67	1,193	6.34	993
10	9.39	2,298	7.54	1,846
12	10.75	3,789	8.65	3,049
15	12.47	6,883	10.10	5,576
18	14.39	11,443	11.68	9,288
21	16.12	17,433	13.13	14,200
24	17.40	24,599	14.23	20,117
27	19.22	34,386	15.78	28,231
30	20.37	44,996	16.71	36,911
33	22.08	59,019	18.16	48,541
36	23.33	74,206	19.21	61,103

TABLE 6-6   Velocity and Discharge in Sewers and Drainage Pipes
(Based on Kutter's Formula, Pipes Flowing Full) *(Continued)*

Diam. of Pipe in In.	N — .011 Velocity in feet per second	N — .011 Discharge gals. per minute	N — .013 Velocity in feet per second	N — .013 Discharge gals. per minute
SLOPE — 5.0 FEET PER HUNDRED FEET				
6	7.09	626	5.64	497
8	8.64	1,344	7.09	1,111
10	10.42	2,550	8.43	2,064
12	12.03	4,241	9.70	3,419
15	13.60	7,507	11.02	6,083
18	16.15	12,843	13.12	10,433
21	18.31	19,802	14.91	16,125
24	19.89	28,119	16.27	23,001
27	21.79	38,984	17.88	31,988
30	22.86	50,496	18.74	41,396
33	24.63	65,835	20.29	54,235
36	26.03	82,796	21.44	68,196

Diam. of Pipe in In.	N — .011 Velocity in feet per second	N — .011 Discharge gals. per minute	N — .013 Velocity in feet per second	N — .013 Discharge gals. per minute
SLOPE — 6.0 FEET PER HUNDRED FEET				
6	7.68	679	6.11	538
8	9.50	1,490	7.68	1,203
10	11.45	2,802	9.13	2,235
12	13.22	4,660	10.51	3,705
15	15.42	8,511	12.49	6,895
18	17.68	14,059	14.36	11,419
21	19.83	21,446	16.12	17,433
24	21.63	30,579	17.59	24,867
27	23.58	42,186	19.36	34,636
30	25.33	55,952	20.77	45,880
33	27.02	72,224	22.26	59,500
36	28.59	90,939	23.54	74,876

Diam. of Pipe in In.	N — .011 Velocity in feet per second	N — .011 Discharge gals. per minute	N — .013 Velocity in feet per second	N — .013 Discharge gals. per minute
SLOPE — 7.0 FEET PER HUNDRED FEET				
6	8.27	731	6.58	580
8	10.26	1,596	8.27	1,296
10	12.38	3,029	9.83	2,406
12	14.18	4,998	11.32	3,990
15	17.00	9,384	13.78	7,607
18	19.04	15,141	15.47	12,302
21	21.37	23,111	17.40	18,818
24	23.38	33,053	19.00	26,861
27	25.38	45,406	20.83	37,266
30	27.29	60,282	22.38	49,436
33	29.15	77,917	24.02	64,205
36	31.02	98,668	25.54	81,238

Diam. of Pipe in In.	N — .011 Velocity in feet per second	N — .011 Discharge gals. per minute	N — .013 Velocity in feet per second	N — .013 Discharge gals. per minute
SLOPE — 8.0 FEET PER HUNDRED FEET				
6	8.86	783	7.05	621
8	11.03	1,715	8.86	1,388
10	13.30	3,255	10.53	2,578
12	15.15	5,340	12.12	4,272
15	18.13	10,007	14.70	8,115
18	20.40	16,222	16.57	13,177
21	22.95	24,820	18.70	20,224
24	25.00	35,343	20.34	28,755
27	27.17	48,609	22.30	39,896
30	29.12	64,324	23.88	52,749
33	31.14	83,237	25.66	68,589
36	33.01	104,998	27.20	86,518

SOURCE: National Clay Pipe Institute.

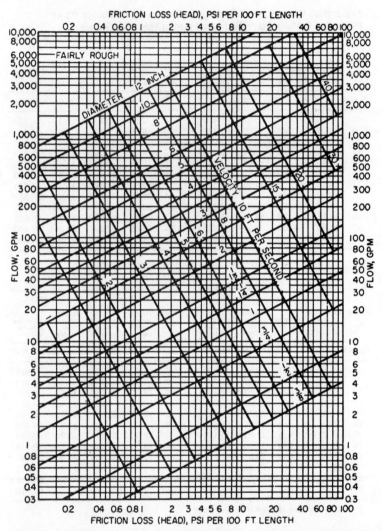

FIGURE 6-14 Friction loss in water piping. *(Courtesy: Standard Handbook of Engineering Calculations by Tyler G. Hicks. Copyright 1972, McGraw-Hill, Inc.)*

TABLE 6-7 Theoretical Discharge of Nozzles, in Gallons per Minute

Head Lbs.	Head Feet	Velocity of discharge, feet per second	1/16	1/8	3/16	1/4	5/16	3/8	7/16	1/2	9/16	5/8	11/16	3/4	13/16	7/8	15/16	1	1 1/8	1 1/4	1 3/8	1 1/2	1 3/4	2	2 1/4	2 1/2	2 3/4	3	3 1/4	3 1/2	3 3/4	4	4 1/2	5	5 1/2	6
10	23.1	38.58	0.37	1.48	3.30	5.90	9.22	13.2	18.1	23.6	29.9	36.8	44.6	53.2	62.3	72.3	83.0	94.4	119	148	178	212	289	378	478	590	715	850	997	1156	1328	1512	1912	2360	2856	3398
15	34.7	47.55	0.45	1.81	4.02	7.23	11.4	16.2	22.3	28.7	36.8	45.0	55.0	65.1	76.8	88.1	102	116	146	181	218	260	354	463	586	728	880	1040	1229	1426	1637	1863	2344	2900	3520	4189
20	46.2	54.55	0.52	2.09	4.66	8.35	13.0	18.7	25.6	33.4	42.2	52.0	63.1	75.3	88.1	102	117	133	169	207	252	300	403	534	676	834	1018	1200	1410	1635	1877	2136	2700	3340	4038	4805
25	57.8	60.99	0.58	2.33	5.23	9.33	14.6	20.9	28.6	37.2	47.2	58.2	70.6	84.1	98.6	114	131	149	189	233	282	336	457	597	756	933	1138	1350	1577	1829	2100	2390	3024	3730	4515	5374
30	69.3	66.82	0.64	2.56	5.71	10.2	16.0	22.8	31.3	40.9	51.7	63.7	77.3	92.2	108	125	144	164	207	256	309	368	501	654	833	1022	1240	1480	1727	2003	2299	2616	3312	4088	4946	5886
35	80.9	72.16	0.69	2.76	6.16	11.0	17.2	24.7	33.8	44.2	55.9	68.9	83.5	99.6	117	135	155	177	223	276	334	397	541	707	895	1104	1340	1590	1865	2163	2483	2828	3580	4415	5341	6357
40	92.4	77.14	0.74	2.95	6.60	11.8	18.4	26.4	36.1	47.2	59.7	73.6	89.2	106	125	144	166	189	239	295	357	425	578	755	956	1180	1430	1700	1994	2313	2655	3020	3824	4720	5710	6796
45	104.0	81.83	0.78	3.13	6.99	12.5	19.6	28.0	38.3	50.2	63.4	78.1	94.7	113	132	153	176	200	253	313	378	450	613	801	1014	1252	1520	1800	2116	2454	2817	3205	4056	5008	6057	7208
50	115.5	86.26	0.83	3.30	7.37	13.2	20.6	29.5	40.4	52.8	66.8	82.3	99.8	119	139	161	186	211	267	330	399	475	646	845	1069	1319	1600	1900	2230	2586	2969	3378	4276	5278	6385	7599
55	127.1	90.46	0.86	3.46	7.73	13.8	21.6	30.9	42.4	55.4	70.1	86.3	104	125	146	169	195	221	280	346	418	498	678	886	1122	1384	1680	2000	2339	2712	3113	3542	4488	5535	6696	7969
60	138.6	94.49	0.90	3.62	8.06	14.5	22.6	32.3	44.3	57.8	73.2	90.1	109	130	153	177	203	231	293	362	437	520	708	925	1171	1446	1755	2090	2443	2834	3253	3701	4694	5784	6994	8324
65	150.2	98.35	0.94	3.77	8.40	15.1	23.5	33.6	46.1	60.2	76.2	93.8	114	136	159	184	212	241	305	377	455	543	737	963	1219	1504	1830	2163	2543	2949	3385	3851	4876	6024	7280	8663
70	161.7	102.06	0.97	3.91	8.73	15.6	24.4	34.9	47.8	62.5	79.0	97.4	118	141	165	191	220	250	316	391	472	563	765	999	1265	1561	1895	2250	2638	3059	3511	3995	5056	6244	7554	8990
75	173.3	105.65	1.01	4.04	9.03	16.1	25.2	36.1	49.5	64.6	81.8	101	122	146	171	198	227	259	327	404	488	583	792	1034	1309	1616	1960	2330	2731	3167	3635	4136	5235	6464	7820	9307
80	184.8	109.11	1.04	4.18	9.33	16.7	26.1	37.5	51.1	66.6	84.5	104	126	150	176	204	235	267	338	417	504	601	818	1068	1353	1669	2030	2404	2820	3271	3755	4272	5407	6676	8076	9612
85	196.4	112.46	1.07	4.31	9.63	17.2	26.9	38.7	52.7	68.8	87.1	107	130	155	182	210	242	275	348	430	520	620	843	1101	1394	1720	2090	2480	2907	3371	3870	4400	5573	6880	8324	9907
90	207.9	115.72	1.10	4.43	9.89	17.7	27.7	39.8	54.2	70.8	89.6	110	134	160	187	217	249	283	358	443	535	637	867	1133	1434	1770	2150	2549	2992	3469	3983	4533	5735	7080	8566	10194
95	219.5	118.89	1.13	4.55	10.2	18.2	28.4	40.9	55.7	72.8	92.0	113	138	164	192	223	256	291	368	455	550	655	891	1164	1474	1818	2200	2618	3073	3564	4091	4656	5891	7280	8801	10473
100	231.0	121.98	1.16	4.67	10.4	18.7	29.2	42.0	57.1	74.6	94.4	116	141	168	197	228	262	298	378	467	564	672	914	1194	1512	1866	2260	2687	3153	3657	4198	4776	6044	7464	9029	10746
105	242.6	125.00	1.19	4.78	10.7	19.1	29.9	43.0	58.5	76.5	96.8	119	145	173	202	234	269	306	387	478	578	688	937	1224	1549	1912	2320	2753	3231	3747	4302	4894	6200	7648	9252	11011
110	254.3	127.94	1.23	4.90	10.9	19.6	30.6	44.0	59.9	78.3	99.1	122	148	177	207	239	275	313	396	489	592	705	959	1253	1586	1957	2380	2818	3307	3835	4403	5010	6340	7828	9470	11270
115	265.7	130.82	1.25	5.01	11.2	20.0	31.3	45.0	61.3	80.1	101	125	151	181	211	245	281	320	405	500	605	720	980	1281	1621	2001	2430	2881	3381	3921	4501	5122	6482	8008	9683	11524
120	277.3	133.63	1.27	5.12	11.4	20.4	31.9	46.0	62.6	81.8	103	127	155	184	216	250	287	327	414	511	618	736	1001	1308	1655	2044	2480	2943	3454	4006	4598	5232	6622	8176	9891	11771
125	288.8	136.38	1.30	5.22	11.7	20.9	32.6	46.9	63.9	83.5	106	130	158	188	220	255	293	334	422	522	631	751	1022	1335	1691	2086	2540	3004	3526	4089	4693	5341	6764	8344	10095	12014
130	300.4	139.08	1.33	5.32	11.9	21.3	33.2	47.9	65.2	85.1	108	133	161	192	225	260	299	340	431	532	644	766	1042	1362	1724	2128	2580	3064	3596	4170	4787	5446	6896	8512	10295	12252

The actual quantity discharged by a nozzle will be less than above table. A well tapered smooth nozzle may be assumed to give above 94 per cent of the values in the tables.

SOURCE: *Pump Handbook* by I. J. Karassik et al. Copyright 1976, McGraw-Hill, Inc.

TABLE 6-8 Pressure Drop of Water through Schedule 40 Steel Pipe

Based on Saph and Schoder formulas: $\Delta P = LQ^{1.86}/1435\,d^5$

The table gives **Pressure Drop of Water per 100 Ft. of Schedule 40 Steel Pipe in psi**. For each pipe size the columns show velocity (v, Ft/Sec) and pressure drop (D, psi). The pipe‑size headings cascade diagonally across the sheet: the seven column‑pairs successively carry 1/8″, 1/4″, 3/8″, 1/2″, 3/4″, 1″, 1 1/4″ (upper band); 1 1/2″, 2″, 2 1/2″, 3″, 3 1/2″, 4″, 5″ (middle band); and 6″, 8″, 10″, 12″, 14″, 16″, 18″ (lower band).

G.P.M.	FT³/Sec	v Ft/Sec	D psi	v Ft/Sec	D psi	v Ft/Sec	D psi	v Ft/Sec	D psi	v Ft/Sec	D psi	v Ft/Sec	D psi	v Ft/Sec	D psi
		1/8″		**1/4″**		**3/8″**		**1/2″**		**3/4″**		**1″**		**1 1/4″**	
1	.00022	5.6	6.77												
2	.00045	1.14	2.48												
3	.00067			1.70	5.26	1.24	1.16	.50	2.55						
4	.00089			2.26	9.00	1.24	1.98	.67	4.36	.42	1.36				
5	.00111			2.82	13.58	1.55	3.00	.84	6.56	.53	2.05	.30	.050		
6	.00134			3.38	19.12	1.85	4.22	1.01	9.25	.63	2.90	.36	.071		
8	.00178			4.52	32.62	2.47	7.17	1.34	1.58	.84	4.94	.48	.121	.30	.036
1	.00223					3.09	10.91	1.68	2.39	1.06	7.49	.60	.183	.37	.055
2	.00446					6.18	39.60	3.36	8.68	2.11	2.72	1.20	.665	.74	.199
3	.00668							5.04	18.46	3.17	5.77	1.80	1.41	1.11	.424
4	.00891							6.72	31.55	4.22	9.86	2.40	2.42	1.49	.724
5	.01114									5.28	14.92	3.01	3.64	1.86	1.09
		1 1/2″		**2″**		**2 1/2″**		**3″**		**3 1/2″**		**4″**		**5″**	
6	.01337							6.33	20.95	3.61	5.13	2.23	1.54	1.29	.390
8	.01782	1.26	.304							4.81	8.76	2.97	2.62	1.71	.667
10	.02228	1.58	.466							6.01	13.28	3.71	3.97	2.14	1.01
15	.03342	2.36	.992	1.43	.285							5.57	8.46	3.21	2.14
20	.04456	3.15	1.69	1.91	.486							7.43	14.42	4.28	3.66
25	.05570	3.94	2.54	2.39	.736	2.01	.474							5.36	5.54
30	.06684	4.73	3.64	2.37	1.03	2.35	.566							6.43	7.79
35	.07798	5.51	4.79	3.35	1.37	2.68	.724							7.50	10.38
40	.08912	6.30	6.14	3.82	1.76	3.15	1.10	2.17	.371					8.57	13.28
50	.1114	7.88	9.31	4.78	2.67					**3 1/2″**					
60	.1337	9.45	13.08	5.74	3.75	4.02	1.54	2.61	.520	2.27	.335				
70	.1560			6.70	4.99	4.70	2.05	3.04	.693	2.59	.430				
80	.1782			7.65	6.40	5.37	2.63	3.47	.890	2.92	.535	**4″**			
90	.2005			8.60	7.96	6.04	3.28	3.91	1.10	3.24	.650	2.52	.346		
100	.2228			9.56	9.69	6.71	3.98	4.34	1.34						
125	.2785	**6″**				8.38	6.03	5.43	2.01	4.05	.984	3.15	.523	**5″**	
150	.3342					10.1	8.46	6.52	2.86	4.87	1.38	3.78	.734	2.81	.316
175	.3899	6″				11.7	11.3	7.60	3.81	5.68	1.84	4.41	.976	3.21	.405
200	.4456					13.4	14.4	8.69	4.89	6.49	2.36	5.04	1.75	3.61	.505
225	.5013							9.77	6.09	7.30	2.94	5.67	1.56	4.01	.616
250	.5570	2.78	.245					10.9	7.41	8.11	3.58	6.30	1.90	4.41	.734
275	.6127	3.06	.282	**8″**				11.9	8.84	8.92	4.27	6.93	2.27	4.81	.863
300	.6684	3.33	.344					13.0	10.4	9.73	5.07	7.56	2.67	5.62	1.15
350	.7798	3.89	.457	2.57	.149			15.2	13.8	11.4	6.87	8.82	3.55	6.41	1.47
400	.8912	4.56	.587							13.0	8.58	10.1	4.56		
450	1.003	5.00	.731	2.89	.185					14.6	10.7	11.3	5.66	7.22	1.82
500	1.114	5.55	.807	3.21	.225					16.2	13.0	12.6	6.89	8.02	2.23
550	1.225	6.11	1.07	3.53	.270	**10″**				17.8	15.5	13.9	8.25	8.82	2.67
600	1.337	6.66	1.25	3.85	.316					19.5	18.2	15.1	9.68	9.62	3.13
650	1.449	7.22	1.45	4.17	.367	2.65	.118					16.4	11.2	10.4	3.62
700	1.560	7.78	1.66	4.49	.420	2.85	.135					17.6	12.9	11.2	4.16
750	1.671	8.33	1.89	4.81	.817	3.05	.154	**12″**				18.9	14.7	12.0	4.75
800	1.782	8.89	2.13	5.13	.480	3.26	.173					20.2	16.5	12.8	5.35
850	1.894	9.44	2.38	5.45	.605	3.46	.194					21.4	18.5	13.6	5.98
900	2.005	10.0	2.66	5.77	.627	3.66	.216	2.54	.090			22.7	20.6	14.4	6.65
950	2.117	10.6		6.09	.744	3.87	.238	2.77	.099	**14″**		23.9	22.8	15.2	7.36
1000	2.228	11.1	3.23	6.41	.817	4.07	.262	2.87	.109					16.0	8.10
1100	2.451	12.2	3.85	7.06	.975	4.48	.313	3.15	.130	2.85	.096			17.6	9.66
1200	2.674	13.3	4.53	7.70	1.15	4.80	.368	3.44	.153	3.08	.111			19.2	11.4
1300	2.896	14.4	5.26	8.34	1.33	5.29	.427	3.73	.178			**16″**		20.8	13.2
1400	3.119	15.6	6.01	8.98	1.53	5.70	.490	4.01	.204	3.32	.127			22.4	15.1
1500	3.342	16.7	6.84	9.62	1.74	6.51	.556	4.30	.232	3.56	.145	2.91	.084	24.1	17.2
1600	3.565	17.8	7.73	10.3	1.96	6.51	.628	4.59	.262	3.79	.163	3.27	.104		
1800	4.010	20.0	9.64	11.5	2.46	7.32	.782	5.16	.329	4.27	.203	3.63	.127	**18″**	
2000	4.456	22.2	11.6	12.8	2.97	8.14	.953	5.73	.396						
2500	5.570	27.8	17.8	16.0	4.49	10.2	1.44	7.17	.601	5.93	.374	4.54	.192		
3000	6.684			19.2	6.30	12.2	2.02	8.60	.842	7.11	.525	5.45	.270	4.30	.149
3500	7.798			22.4	8.41	14.2	2.70	10.0	1.12	8.30	.700	6.36	.358	5.02	.199
4000	8.912			25.7	10.8	16.3	3.46	11.5	1.44	9.48	.896	7.26	.456	5.74	.255
4500	10.03			28.9	13.4	18.3	4.31	12.9	1.76	10.7	1.12	8.17	.671	6.45	.317
5000	11.14					20.4	5.20	14.3	2.18	11.9	1.36	9.08	.695	7.17	.386
6000	13.37					24.4	7.35	17.2	3.06	14.2	1.91	10.9	.977	8.60	.542
7000	15.60					28.5	9.80	20.1	4.08	16.6	2.54	14.5	1.30	10.0	.723
8000	17.82							22.9	5.22	19.0	3.25	14.5	1.67	11.5	.926
9000	20.05							25.8	6.51	21.3	4.06	16.3	2.08	12.9	1.15
10000	22.28							28.7	7.91	23.7	4.97	18.2	2.53	14.3	1.40
12000	26.74									28.5	6.92	21.8	3.55	17.2	1.97
14000	31.19											25.4	4.72	20.1	2.62
16000	35.65											29.1	6.06	22.9	3.36
18000	40.10											32.7	7.55	25.8	4.18
20000	44.56													28.7	5.08

SOURCE: *Fluid Flow Pocket Handbook* by Nicholas P. Cheremisinoff. Copyright 1984, Gulf Publishing Co., Houston, Texas. Used with permission. All rights reserved.

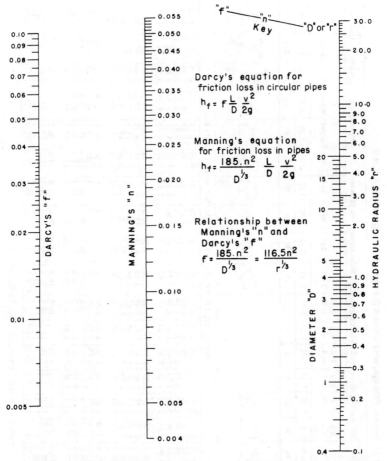

FIGURE 6-15 Relationship between Darcy's *f* and Manning's *n* for flow in pipes. *(Courtesy: Design of Small Dams, U.S. Department of Interior, Bureau of Reclamation.)*

TABLE 6-9 Equivalent Resistance of Bends, Fittings, and Valves (Length of Straight Pipe in Feet)

Nominal pipe size, in.	Inside diam. d, in. Sched. 40	Screwed fittings				90° welding elbows and smooth bends						Miter elbows (No. of miters)					Welding tees		Valves (screwed, flanged, or welded)			
		45° ell	90° ell	180° close return bends	Tee	R/d =1	R/d =1½	R/d =2	R/d =4	R/d =6	R/d =8	1-45°	1-60°	1-90°	2-90°	3-90°	Forged	Miter	Gate	Globe	Angle	Swing check
k factor =		0.42	0.90	2.00	1.80	0.48	0.36	0.27	0.21	0.27	0.36	0.45	0.90	1.80	0.60	0.45	1.35	1.80	0.21	10	5.0	2.5
L/d' ratio n =		14	30	67	60	16	12	9	7	9	12	15	30	60	20	15	45	60	7	333	167	83

L = equivalent length in feet of Schedule 40 (standard weight) straight pipe

Nominal pipe size, in.	Inside diam. d, in. Sched. 40	45° ell	90° ell	180° close return bends	Tee	R/d=1	R/d=1½	R/d=2	R/d=4	R/d=6	R/d=8	1-45°	1-60°	1-90°	2-90°	3-90°	Forged	Miter	Gate	Globe	Angle	Swing check
½	0.622	0.73	1.55	3.47	3.10	0.83	0.62	0.47	0.36	0.47	0.62	0.78	1.55	3.10	1.04	0.78	2.33	3.10	0.36	17.3	8.65	4.32
¾	0.824	0.96	2.06	4.60	4.12	1.10	0.82	0.62	0.48	0.62	0.82	1.03	2.06	4.12	1.37	1.03	3.09	4.12	0.48	22.9	11.4	5.72
1	1.049	1.22	2.62	5.82	5.24	1.40	1.05	0.79	0.61	0.79	1.05	1.31	2.62	5.24	1.75	1.31	3.93	5.24	0.61	29.1	14.6	7.27
1¼	1.380	1.61	3.45	7.66	6.90	1.84	1.38	1.03	0.81	1.03	1.38	1.72	3.45	6.90	2.30	1.72	5.17	6.90	0.81	38.3	19.1	9.58
1½	1.610	1.88	4.02	8.95	8.04	2.15	1.61	1.21	0.94	1.21	1.61	2.01	4.02	8.04	2.68	2.01	6.04	8.04	0.94	44.7	22.4	11.2
2	2.067	2.41	5.17	11.5	10.3	2.76	2.07	1.55	1.21	1.55	2.07	2.58	5.17	10.3	3.45	2.58	7.75	10.3	1.21	57.4	28.7	14.4
2½	2.469	2.88	6.16	13.7	12.3	3.29	2.47	1.85	1.44	1.85	2.47	3.08	6.16	12.3	4.11	3.08	9.25	12.3	1.44	68.5	34.3	17.1
3	3.068	3.58	7.67	17.1	15.3	4.09	3.07	2.30	1.79	2.30	3.07	3.84	7.67	15.3	5.11	3.84	11.5	15.3	1.79	85.2	42.6	21.3
4	4.026	4.70	10.1	22.4	20.2	5.37	4.03	3.02	2.35	3.02	4.03	5.04	10.1	20.2	6.71	5.04	15.1	20.2	2.35	112	56.0	28.0
5	5.047	5.88	12.6	28.0	25.2	6.72	5.05	3.78	2.94	3.78	5.05	6.30	12.6	25.2	8.40	6.30	18.9	25.2	2.94	140	70.0	35.0
6	6.065	7.07	15.2	33.8	30.4	8.09	6.07	4.55	3.54	4.55	6.07	7.58	15.2	30.4	10.1	7.58	22.8	30.4	3.54	168	84.1	42.1
8	7.981	9.31	20.0	44.6	40.0	10.6	7.98	5.98	4.65	5.98	7.98	9.97	20.0	40.0	13.3	9.97	29.8	40.0	4.65	222	111	55.5
10	10.02	11.7	25.0	55.7	50.0	13.3	10.0	7.51	5.85	7.51	10.0	12.5	25.0	50.0	16.7	12.5	37.6	50.0	5.85	278	139	69.5
12	11.94	13.9	29.8	66.3	59.6	15.9	11.9	8.95	6.96	8.95	11.9	14.9	29.8	59.6	19.9	14.9	44.8	59.6	6.96	332	166	83.0
14	13.13	15.3	32.8	73.0	65.6	17.5	13.1	9.85	7.65	9.85	13.1	16.4	32.8	65.6	21.9	16.4	49.2	65.6	7.65	364	182	91.0
16	15.00	17.5	37.5	83.5	75.0	20.0	15.0	11.2	8.75	11.2	15.0	18.8	37.5	75.0	25.0	18.8	56.2	75.0	8.75	417	208	104
18	16.88	19.7	42.0	93.8	84.2	22.5	16.9	12.7	9.85	12.7	16.9	21.1	42.1	84.2	28.1	21.1	63.2	84.2	9.85	469	234	117
20	18.81	22.0	47.0	105	94.0	25.1	18.8	14.1	11.0	14.1	18.8	23.5	47.0	94.0	31.4	23.5	70.6	94.0	11.0	522	261	131
24	22.63	26.4	56.6	126	113	30.2	22.6	17.0	13.2	17.0	22.6	28.3	56.6	113	37.8	28.3	85.0	113	13.2	629	314	157

SOURCE: Piping Handbook, 4th ed., by Sabin Crocker. Copyright 1945, McGraw-Hill, Inc.

NOTES

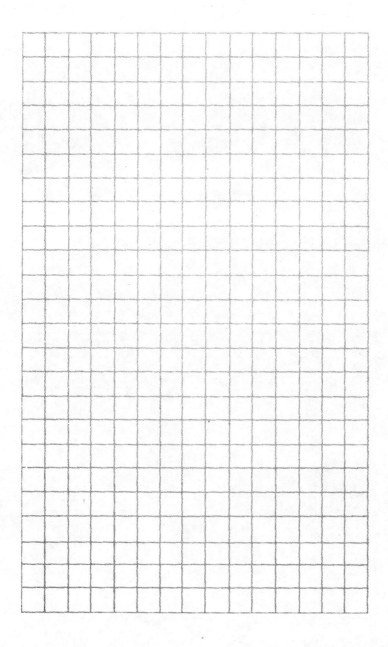

Culverts and Storm Water

Shape		Range of Sizes	Common Uses
Round		6 in.–26 ft	Culverts, subdrains, sewers, service tunnels, etc. All plates same radius. For medium and high fills (or trenches).
Vertically-elongated (ellipse) 5% is common		4–21 ft nominal; before elongating	Culverts, sewers, service tunnels, recovery tunnels. Plates of varying radii; shop fabrication. For appearance and where backfill compaction is only moderate.
Pipe-arch		Span x Rise 18 in. x 11 in. to 20 ft 7 in. x 13 ft 2 in.	Where headroom is limited. Has hydraulic advantages at low flows. Corner plate radius, 18 inches or 31 inches for structural plate.
Underpass*		Span x Rise 5 ft 8 in. x 5 ft 9 in. to 20 ft 4 in. x 17 ft 9 in.	For pedestrians, livestock or vehicles (structural plate).
Arch		Span x Rise 6 ft x 1 ft 9½ in. to 25 ft x 12 ft 6 in.	For low clearance large waterway opening, and aesthetics (structural plate).
Horizontal Ellipse		Span 20–40 ft	Culverts, grade separations, storm sewers, tunnels.
Pear		Span 25–30 ft	Grade separations, culverts, storm sewers, tunnels.
High Profile Arch		Span 20–45 ft	Culverts, grade separations, storm sewers, tunnels, Ammo ammunition magazines, earth covered storage.
Low Profile Arch		Span 20–50 ft	Low-Wide waterway enclosures, culverts, storm sewers.
Box Culverts		Span 10–21 ft	Low-wide waterway enclosures, culverts, storm sewers.
Specials		Various	For lining old structures or other special purposes. Special fabrication.

*For equal area or clearance, the round shape is generally more economical and simpler to assemble.

FIGURE 7-1 Typical shapes and uses of corrugated conduits. *(Courtesy: American Iron and Steel Institute.)*

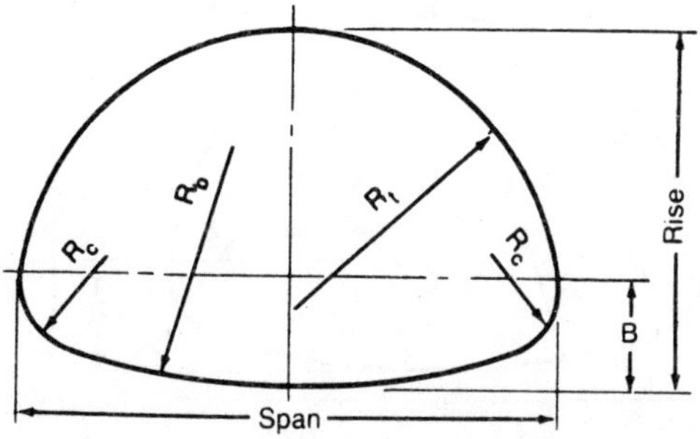

Equiv. Diameter, in.	Span, in.	Rise, in.	Waterway Area, ft²	Layout Dimensions			
				B in.	R_c in.	R_t in.	R_b in.
15	17	13	1.1	4⅛	3½	8⅝	25⅝
18	21	15	1.6	4⅞	4⅛	10¾	33⅛
21	24	18	2.2	5⅝	4⅞	11⅞	34⅝
24	28	20	2.9	6½	5½	14	42¼
30	35	24	4.5	8⅛	6⅞	17⅞	55⅛
36	42	29	6.5	9¾	8¼	21½	66⅛
42	49	33	8.9	11⅜	9⅝	25⅛	77¼
48	57	38	11.6	13	11	28⅝	88¼
54	64	43	14.7	14⅝	12⅜	32¼	99¼
60	71	47	18.1	16¼	13¾	35¾	110¼
66	77	52	21.9	17⅞	15⅛	39⅜	121¼
72	83	57	26.0	19½	16½	43	132¼

Dimensions shown not for specification purposes, subject to manufacturing tolerances.

(a)

Equiv. Diameter, in.	Size, in.	Span, in.	Rise, in.	Waterway Area, ft²	Layout Dimensions			
					B in.	R_c in.	R_t in.	R_b in.
54	60 × 46	58½	48½	15.6	20½	18¾	29⅜	51⅛
60	66 × 51	65	54	19.3	22¾	20¾	32⅝	56¼
66	73 × 55	72½	58¼	23.2	25⅛	22⅞	36¾	63¾
72	81 × 59	79	62½	27.4	23¾	20⅞	39½	82⅝
78	87 × 63	86½	67¼	32.1	25¾	22⅝	43⅜	92¼
84	95 × 67	93½	71¾	37.0	27¾	24⅜	47	100¼
90	103 × 71	101½	76	42.4	29¾	26⅛	51¼	111⅝
96	112 × 75	108½	80½	48.0	31⅝	27¾	54⅞	120¼
102	117 × 79	116½	84¾	54.2	33⅝	29½	59⅜	131¾
108	128 × 83	123½	89¼	60.5	35⅝	31¼	63¼	139¾
114	137 × 87	131	93¾	67.4	37⅝	33	67⅜	149½
120	142 × 91	138½	98	74.5	39½	34¾	71⅝	162⅜

Dimensions shown not for specification purposes, subject to manufacturing tolerances.

(b)

FIGURE 7-2 Sizes and layout details—CSP pipe arches. (*a*) 2⅔-inch × ½-inch corrugation. (*b*) 3-inch × 1-inch corrugation. *(Courtesy: American Iron and Steel Institute.)*

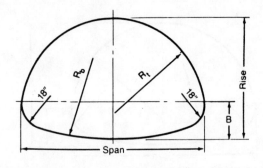

Dimensions			Layout Dimensions			Periphery		
Span, ft-in.	Rise, ft-in.	Waterway Area, ft²	B in.	R_f ft	R_b ft	No. of Plates	Total	
							N	Pi
6-1	4-7	22	21.0	3.07	6.36	5	22	66
6-4	4-9	24	20.5	3.18	8.22	5	23	69
6-9	4-11	26	22.0	3.42	6.96	5	24	72
7-0	5-1	28	21.4	3.53	8.68	5	25	75
7-3	5-3	31	20.8	3.63	11.35	6	26	78
7-8	5-5	33	22.4	3.88	9.15	6	27	81
7-11	5-7	35	21.7	3.98	11.49	6	28	84
8-2	5-9	38	20.9	4.08	15.24	6	29	87
8-7	5-11	40	22.7	4.33	11.75	7	30	90
8-10	6-1	43	21.8	4.42	14.89	7	31	93
9-4	6-3	46	23.8	4.68	12.05	7	32	96
9-6	6-5	49	22.9	4.78	14.79	7	33	99
9-9	6-7	52	21.9	4.86	18.98	7	34	102
10-3	6-9	55	23.9	5.13	14.86	7	35	105
10-8	6-11	58	26.1	5.41	12.77	7	36	108
10-11	7-1	61	25.1	5.49	15.03	7	37	111
11-5	7-3	64	27.4	5.78	13.16	7	38	114
11-7	7-5	67	26.3	5.85	15.27	8	39	117
11-10	7-7	71	25.2	5.93	18.03	8	40	120
12-4	7-9	74	27.5	6.23	15.54	8	41	123
12-6	7-11	78	26.4	6.29	18.07	8	42	126
12-8	8-1	81	25.2	6.37	21.45	8	43	129
12-10	8-4	85	24.0	6.44	26.23	8	44	132
13-5	8-5	89	26.3	6.73	21.23	9	45	135
13-11	8-7	93	28.9	7.03	18.39	9	46	138
14-1	8-9	97	27.6	7.09	21.18	9	47	141
14-3	8-11	101	26.3	7.16	24.80	9	48	144
14-10	9-1	105	28.9	7.47	21.19	9	49	147
15-4	9-3	109	31.6	7.78	18.90	9	50	150
15-6	9-5	113	30.2	7.83	21.31	10	51	153
15-8	9-7	118	28.8	7.89	24.29	10	52	156
15-10	9-10	122	27.4	7.96	28.18	10	53	159
16-5	9-11	126	30.1	8.27	24.24	10	54	162
16-7	10-1	131	28.7	8.33	27.73	10	55	165

Dimensions are to inside crests and are subject to manufacturing tolerances.
$N = 3$ Pi $= 9.6$ in.

FIGURE 7-3 Sizes and layout details—structural plate steel pipe arches. 18-inch corner radius; 6-inch × 2-inch corrugations—bolted seams. *(Courtesy: American Iron and Steel Institute.)*

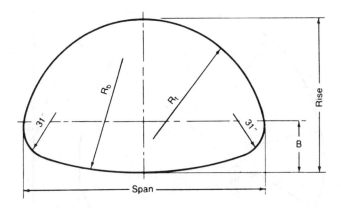

Dimensions			Layout Dimensions			Periphery		
		Waterway	B	R_t	R_b	No. of	Total	
Span, ft-in.	Rise, ft-in.	Area, ft²	in.	ft	ft	Plates	N	Pi
13-3	9-4	97	38.5	6.68	16.05	8	46	138
13-6	9-6	102	37.7	6.78	18.33	8	47	141
14-0	9-8	105	39.6	7.03	16.49	8	48	144
14-2	9-10	109	38.8	7.13	18.55	8	49	147
14-5	10-0	114	37.9	7.22	21.38	8	50	150
14-11	10-2	118	39.8	7.48	18.98	9	51	153
15-4	10-4	123	41.8	7.76	17.38	9	52	156
15-7	10-6	127	40.9	7.84	19.34	10	53	159
15-10	10-8	132	40.0	7.93	21.72	10	54	162
16-3	10-10	137	42.1	8.21	19.67	10	55	165
16-6	11-0	142	41.1	8.29	21.93	10	56	168
17-0	11-2	146	43.3	8.58	20.08	10	57	171
17-2	11-4	151	42.3	8.65	22.23	10	58	174
17-5	11-6	157	41.3	8.73	24.83	10	59	177
17-11	11-8	161	43.5	9.02	22.55	10	60	180
18-1	11-10	167	42.4	9.09	24.98	10	61	183
18-7	12-0	172	44.7	9.38	22.88	10	62	186
18-9	12-2	177	43.6	9.46	25.19	10	63	189
19-3	12-4	182	45.9	9.75	23.22	10	64	192
19-6	12-6	188	44.8	9.83	25.43	11	65	195
19-8	12-8	194	43.7	9.90	28.04	11	66	198
19-11	12-10	200	42.5	9.98	31.19	11	67	201
20-5	13-0	205	44.9	10.27	28.18	11	68	204
20-7	13-2	211	43.7	10.33	31.13	12	69	207

Dimensions are to inside crests and are subject to manufacturing tolerances.
$N = 3$ Pi $= 9.6$ in.

FIGURE 7-4 Sizes and layout details — structural plate steel pipe arches.
31-inch corner radius; 6-inch × 2-inch corrugations — bolted seams. (*Courtesy: American Iron and Steel Institute.*)

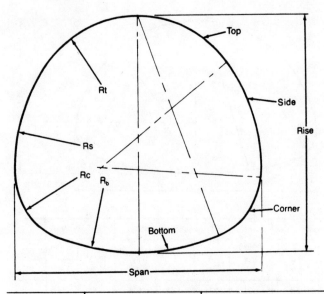

	Periphery			Layout Dimensions in In.			
Span × Rise, ft and in.	N	Pi	No. of Plates per Ring	R_t	R_s	R_c	R_b
5-8 5-9	24	72	6	27	53	18	Flat
5-8 6-6	26	78	6	29	75	18	Flat
5-9 7-4	28	84	6	28	95	18	Flat
5-10 7-8	29	87	7	30	112	18	Flat
5-10 8-2	30	90	6	28	116	18	Flat
12-2 11-0	47	141	8	68	93	38	136
12-11 11-2	49	147	9	74	92	38	148
13-2 11-10	51	153	11	73	102	38	161
13-10 12-2	53	159	11	77	106	38	168
14-1 12-10	55	165	11	77	115	38	183
14-6 13-5	57	171	11	78	131	38	174
14-10 14-0	59	177	11	79	136	38	193
15-6 14-4	61	183	12	83	139	38	201
15-8 15-0	63	189	12	82	151	38	212
16-4 15-5	65	195	12	86	156	38	217
16-5 16-0	67	201	12	88	159	38	271
16-9 16-3	68	204	12	89	168	38	246
17-3 17-0	70	210	12	90	174	47	214
18-4 16-11	72	216	12	99	157	47	248
19-1 17-2	74	222	13	105	156	47	262
19-6 17-7	76	228	13	107	158	47	295
20-4 17-9	78	234	13	114	155	47	316

All dimensions, to nearest whole number, are measured from inside crests.
Tolerances should be allowed for specification purposes.   6 × 2 in. Corrugations.

FIGURE 7-5 Structural plate steel underpasses—sizes and layout details. *(Courtesy: American Iron and Steel Institute.)*

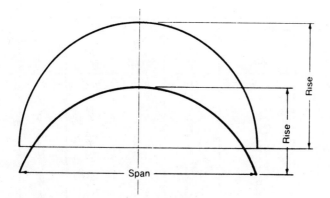

Dimensions[1]		Waterway Area, ft²	Rise over Span[2]	Radius, in.	Nominal Arc Length	
Span, ft	Rise, ft-in.				N[3]	Pi, in.
6.0	1-9½	7½	0.30	41	9	27
	2-3½	10	0.38	37½	10	30
	3-2	15	0.53	36	12	36
7.0	2-4	12	0.34	45	11	33
	2-10	15	0.40	43	12	36
	3-8	20	0.52	42	14	42
8.0	2-11	17	0.37	51	13	39
	3-4	20	0.42	48½	14	42
	4-2	26	0.52	48	16	48
9.0	2-11	18½	0.32	59	14	42
	3-10½	26½	0.43	55	16	48
	4-8½	33	0.52	54	18	54
10.0	3-5½	25	0.35	64	16	48
	4-5	34	0.44	60½	18	54
	5-3	41	0.52	60	20	60
11.0	3-6	27½	0.32	73	17	51
	4-5½	37	0.41	67½	19	57
	5-9	50	0.52	66	22	66
12.0	4-0½	35	0.34	77½	19	57
	5-0	45	0.42	73	21	63
	6-3	59	0.52	72	24	72
13.0	4-1	38	0.32	86½	20	60
	5-1	49	0.39	80½	22	66
	6-9	70	0.52	78	26	78
14.0	4-7½	47	0.33	91	22	66
	5-7	58	0.40	86	24	72
	7-3	80	0.52	84	28	84

[1]Dimensions are to inside crests and are subject to manufacturing tolerances.
[2]R/S ratio varies from 0.30 to 0.52. Intermediate spans and rises are available.
[3]W = 3 Pi = 9.6 in.   6 × 2 in. Corrugations—Bolted Seams.

FIGURE 7-6 Representative sizes of structural plate steel arches.
(Courtesy: American Iron and Steel Institute.)

Dimensions[1]		Waterway Area, ft²	Rise over Span[2]	Radius, in.	Nominal Arc Length	
Span, ft	Rise, ft-in.				N[3]	Pi, in.
15.0	4-7½	50	0.31	101	23	69
	5-8	62	0.38	93	25	75
	6-7	75	0.44	91	27	81
	7-9	92	0.52	90	30	90
16.0	5-2	60	0.32	105	25	75
	7-1	86	0.45	97	29	87
	8-3	105	0.52	96	32	96
17.0	5-2½	63	0.31	115	26	78
	7-2	92	0.42	103	30	90
	8-10	119	0.52	102	34	102
18.0	5-9	75	0.32	119	28	84
	7-8	104	0.43	109	32	96
	8-11	126	0.50	108	35	105
19.0	6-4	87	0.33	123	30	90
	8-2	118	0.43	115	34	102
	9-5½	140	0.50	114	37	111
20.0	6-4	91	0.32	133	31	93
	8-3½	124	0.42	122	35	105
	10-0	157	0.50	120	39	117
21.0	6-11	104	0.33	137	33	99
	8-10	140	0.42	128	37	111
	10-6	172	0.50	126	41	123
22.0	6-11	109	0.31	146	34	102
	8-11	146	0.40	135	38	114
	11-0	190	0.50	132	43	129
23.0	8-0	134	0.35	147	37	111
	9-10	171	0.43	140	41	123
	11-6	208	0.50	138	45	135
24.0	8-6	150	0.35	152	39	117
	10-4	188	0.43	146	43	129
	12-0	226	0.50	144	47	141
25.0	8-6½	155	0.34	160	40	120
	10-10½	207	0.43	152	45	135
	12-6	247	0.50	150	49	147

[1]Dimensions are to inside crests and are subject to manufacturing tolerances.
[2]R/S ratio varies from 0.30 to 0.52. Intermediate spans and rises are available.
[3]N = 3 Pi = 9.6 in.   6 × 2 in. Corrugations—Bolted Seams.

FIGURE 7-6 *(Continued)*

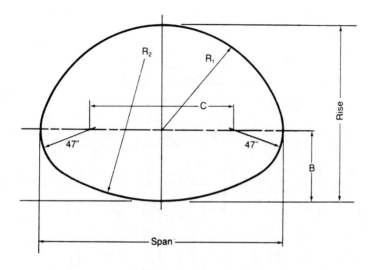

Span, ft-in.	Rise, ft-in.	Area, ft²	Total No. Plates	Periphery						B, in.	C, in.	Inside Radius	
				Top		Bottom		Total				R_1, in.	R_2, in.
				N	Pi	N	Pi	N	Pi				
20- 0	13-11	218	10	34	102	20	60	68	204	62.8	146.2	122.5	223.6
20- 6	14- 3	231	10	36	108	20	60	70	210	61.4	152.3	124.7	255.7
21- 5	14- 6	243	11	36	108	22	66	72	216	65.3	162.8	131.4	236.7
21-11	14-11	256	11	38	114	22	66	74	222	63.7	168.9	133.5	268.1
22- 5	15- 3	270	11	40	120	22	66	76	228	62.1	174.6	135.5	307.1
23- 4	15- 7	284	11	40	120	24	72	78	234	66.2	185.5	142.4	280.2
24- 2	15-11	297	12	40	120	26	78	80	240	70.7	196.2	149.7	262.1
24- 8	16- 2	312	12	42	126	26	78	82	246	68.8	202.2	151.4	292.2
25- 2	16- 7	326	12	44	132	26	78	84	252	66.9	207.9	153.2	328.6
25- 7	16-11	342	12	46	138	26	78	86	258	64.8	213.3	155.0	373.3
26- 7	17- 3	357	12	46	138	28	84	88	264	69.4	224.7	162.1	339.4
27- 6	17- 6	372	12	46	138	30	90	90	270	74.2	235.8	169.6	315.8
28- 0	17-10	388	12	48	144	30	90	92	276	72.1	241.5	171.1	350.2
28- 5	18- 3	405	13	50	150	30	90	94	282	69.9	246.8	172.7	392.3
29- 4	18- 6	421	13	50	150	32	96	96	288	74.8	258.2	180.2	361.1
30- 4	18-10	438	14	52	156	34	102	100	300	80.0	269.4	188.2	339.1

*Includes 14N for two N7 corner plates.

FIGURE 7-7 Long-span pipe arch sizes and layout details. *(Courtesy: American Iron and Steel Institute.)*

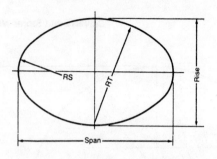

Span, ft-in.	Rise, ft-in.	Area, ft²	Periphery						Inside Radius	
			Top or Bottom		Side		Total		Top	Side
			N	Pi	N	Pi	N	Pi	Rad. in.	Rad. in.
19- 4	12- 9	191	22	66	10	30	64	192	12- 6	4- 6
20- 1	13- 0	202	23	69	10	30	66	198	13- 1	4- 6
20- 2	11-11	183	24	72	8	24	64	192	13- 8	3- 7
20-10	12- 2	194	25	75	8	24	66	198	14- 3	3- 7
21- 0	15- 2	248	23	69	13	39	72	216	13- 1	5-11
21-11	13- 1	221	26	78	9	27	70	210	14-10	4- 1
22- 6	15- 8	274	25	75	13	39	76	228	14- 3	5-11
23- 0	14- 1	249	27	81	10	30	74	222	15- 5	4- 6
23- 3	15-11	288	26	78	13	39	78	234	14-10	5-11
24- 4	16-11	320	27	81	14	42	82	246	15- 5	6- 4
24- 6	14- 8	274	29	87	10	30	78	234	16- 6	4- 6
25- 2	14-11	287	30	90	10	30	80	240	17- 1	4- 6
25- 5	16- 9	330	29	87	13	39	84	252	16- 6	5-11
26- 1	18- 2	369	29	87	15	45	88	264	16- 6	6-10
26- 3	15-10	320	31	93	11	33	84	252	17- 8	4-11
27- 0	16- 2	334	32	96	11	33	86	258	18- 3	4-11
27- 2	19- 1	405	30	90	16	48	92	276	17- 1	7- 3
27-11	19- 5	421	31	92	16	48	94	282	17- 8	7- 3
28- 1	17- 1	369	33	99	12	36	90	270	18-10	5- 5
28-10	17- 5	384	34	102	12	36	92	276	19- 5	5- 5
29- 5	19-11	455	33	99	16	48	98	294	18-10	7- 3
30- 1	20- 2	472	34	102	16	48	100	300	19- 5	7- 3
30- 3	17-11	415	36	108	12	36	96	288	20- 7	5- 5
31- 2	21- 2	512	35	105	17	51	104	312	20- 0	7- 9
31- 4	18-11	454	37	111	13	39	100	300	21- 1	5-11
32- 1	19- 2	471	38	114	13	39	102	306	21- 8	5-11
32- 3	22- 2	555	36	108	18	54	108	324	20- 7	8- 2
33- 0	22- 5	574	37	111	18	54	110	330	21- 1	8- 2
33- 2	20- 1	512	39	117	14	42	106	318	22- 3	6- 4
34- 1	23- 4	619	38	114	19	57	114	342	21- 8	8- 8
34- 7	20- 8	548	41	123	14	42	110	330	23- 5	6- 4
34-11	21- 4	574	41	123	15	45	112	336	23- 5	6-10
35- 1	24- 4	665	39	117	20	60	118	354	22- 3	9- 1
35- 9	25- 9	718	39	117	22	66	122	366	22- 3	10- 0
36- 0	22- 4	619	42	126	16	48	116	348	24- 0	7- 3
36-11	25- 7	735	41	123	21	63	124	372	23- 5	9- 7
37- 2	22- 2	631	44	132	15	45	118	354	25- 2	6-10
38- 0	26- 7	785	44	132	22	66	128	384	24- 0	10- 0
38- 8	27-11	843	42	126	24	72	132	396	24- 0	10-11
40- 0	29- 7	927	43	129	26	78	138	414	27-11	11-10

FIGURE 7-8 Long-span horizontal ellipse pipe sizes and layout details. *(Courtesy: American Iron and Steel Institute.)*

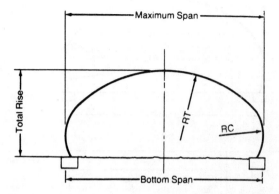

Max. Span, ft-in.	Bottom Span, ft-in.	Total Rise, ft-in.	Area, ft²	Periphery						Inside Radius	
				Top		Side		Total		Top rad. in.	Side rad. in.
				N	Pi	N	Pi	N	Pi		
20- 1	19-10	7- 6	121	23	69	6	18	35	105	13- 1	4- 6
19- 5	19- 1	6-10	105	23	69	5	15	33	99	13- 1	3- 7
21- 6	21- 4	7- 9	134	25	75	6	18	37	111	14- 3	4- 6
22- 3	22- 1	7-11	140	26	78	6	18	38	114	14-10	4- 6
23- 0	22- 9	8- 0	147	27	81	6	18	39	117	15- 5	4- 6
23- 9	23- 6	8- 2	154	28	84	6	18	40	120	16- 0	4- 6
24- 6	24- 3	8- 4	161	29	87	6	18	41	123	16- 6	4- 6
25- 2	25- 0	8- 5	169	30	90	6	18	42	126	17- 1	4- 6
25-11	25- 9	8- 7	176	31	93	6	18	43	129	17- 8	4- 6
27- 3	27- 1	10- 0	217	31	93	8	24	47	141	17- 8	6- 4
28- 1	27-11	9- 7	212	33	99	7	21	47	141	18-10	5- 5
28- 9	28- 7	10- 3	234	33	99	8	24	49	147	18-10	6- 4
28-10	28- 8	9- 8	221	34	102	7	21	48	144	19- 5	5- 5
30- 3	30- 1	9-11	238	36	108	7	21	50	150	20- 7	5- 5
30-11	30- 9	10- 8	261	36	108	8	24	52	156	20- 7	6- 4
31- 7	31- 2	12- 1	309	36	108	10	30	56	168	20- 7	7- 3
31- 0	30-10	10- 1	246	37	111	7	21	51	153	21- 1	5- 5
32- 4	31-11	12- 3	320	37	111	10	30	57	171	21- 1	7- 3
31- 9	31- 7	10- 3	255	38	114	7	21	52	156	21- 8	5- 5
33- 1	32- 7	12- 5	330	38	114	10	30	58	174	21- 8	7- 3
33- 2	33- 0	11- 1	289	39	117	8	24	55	165	22- 3	6- 4
34- 5	34- 1	13- 3	377	39	117	11	33	61	183	22- 3	8- 2
34- 7	34- 6	11- 4	308	41	123	8	24	57	183	23- 5	6- 4
37-11	37- 7	15- 8	477	41	123	14	42	69	207	23- 5	10-11
35- 4	35- 2	11- 5	318	42	126	8	24	58	174	24- 0	6- 4
38- 8	38- 4	15- 9	490	42	126	14	42	70	210	24- 0	10-11

NOTE: Larger sizes available for special designs.

FIGURE 7-9 Long-span low-profile arch pipe sizes and layout details. *(Courtesy: American Iron and Steel Institute.)*

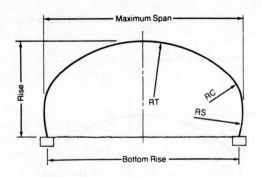

Max. Span, ft-in.	Bottom Span, ft-in.	Total Rise, ft-in.	Area, ft²	Periphery								Inside Radius		
				Top		Upper Side		Lower Side		Total		Top Radius, ft-in.	Upper Side, ft-in.	Lower Side, ft-in.
				N	Pi	N	Pi	N	Pi	N	Pi			
20- 1	19- 6	9- 1	152	23	69	5	15	3	9	39	117	13- 1	4- 6	13- 1
20- 8	18-10	12- 1	214	23	69	6	18	6	18	47	141	13- 1	5- 5	13- 1
21- 6	19-10	11- 8	215	25	75	5	15	6	18	47	141	14- 3	4- 6	14- 3
22-10	19-10	14- 7	285	25	75	7	21	8	24	55	165	14- 3	6- 4	14- 3
22- 3	20- 7	11-10	225	26	78	5	15	6	18	48	144	14-10	4- 6	14-10
22-11	20- 0	14- 0	276	26	78	6	18	8	24	54	162	14-10	5- 5	14-10
23- 0	21- 5	12- 0	235	27	81	5	15	6	18	49	147	15- 5	4- 6	15- 5
24- 4	21- 6	14-10	310	27	81	7	21	8	24	57	171	15- 5	6- 4	15- 5
23- 9	22- 2	12- 1	245	28	84	5	15	6	18	50	150	16- 0	4- 6	16- 0
24- 6	21-11	13- 9	289	29	87	5	15	8	24	55	165	16- 6	4- 6	16- 6
25- 9	23- 2	15- 2	335	29	87	7	21	8	24	59	177	16- 6	6- 4	16- 6
25- 2	23- 3	13- 2	283	30	90	5	15	7	21	54	162	17- 1	4- 6	17- 1
26- 6	24- 0	15- 3	348	30	90	7	21	8	24	60	180	17- 1	6- 4	17- 1
25-11	24- 1	13- 3	295	31	93	5	15	7	21	55	165	17- 8	4- 6	17- 8
27- 3	24-10	15- 5	360	31	93	7	21	8	24	61	183	17- 8	6- 4	17- 8
27- 5	25- 8	13- 7	317	33	99	5	15	7	21	57	171	18-10	4- 6	18-10
29- 5	27- 1	16- 5	412	33	99	8	28	8	24	65	195	18-10	7- 3	18-10
28- 2	25-11	14- 5	349	34	102	5	15	8	24	60	180	19- 5	4- 6	19- 5
30- 1	26- 9	18- 1	467	34	102	8	24	10	30	70	210	19- 5	7- 3	19- 5
30- 3	28- 2	15- 5	399	36	108	6	18	8	24	64	192	20- 7	5- 5	20- 7
31- 7	28- 4	18- 4	497	36	108	8	24	10	30	72	216	20- 7	7- 3	20- 7
31- 0	29- 0	15- 7	413	37	111	6	18	8	24	65	195	21- 1	5- 5	21- 1
31- 8	28- 6	17- 9	484	37	111	7	21	10	30	71	213	21- 1	6- 4	21- 1
32- 4	27-11	19-11	554	37	111	8	24	12	36	77	231	21- 1	7- 3	21- 1
31- 9	28- 8	17- 3	470	38	114	6	18	10	30	70	210	21- 8	5- 5	21- 8
33- 1	28- 9	20- 1	571	38	114	8	24	12	36	78	234	21- 8	7- 3	21- 8
32- 6	29- 6	17- 4	484	39	117	6	18	10	30	71	213	22- 3	5- 5	22- 3
33-10	29- 7	20- 3	588	39	117	8	24	12	36	79	237	22- 3	7- 3	22- 3
34- 0	31- 2	17- 8	514	41	123	6	18	10	30	73	219	23- 5	5- 5	23- 5
34- 7	30- 7	19-10	591	41	123	7	21	12	36	79	237	23- 5	6- 4	23- 5
35- 3	30- 7	21- 3	645	41	123	8	24	13	39	83	249	23- 5	7- 3	23- 5
37- 3	32- 6	23- 5	747	41	123	11	33	13	39	89	267	23- 5	10- 0	23- 5
34- 8	31-11	17-10	529	42	126	6	18	10	30	74	222	24- 0	5- 5	24- 0
35- 4	31- 5	20- 0	608	42	126	7	21	12	36	80	240	24- 0	6- 4	24- 0
36- 0	31- 5	21- 5	663	42	126	8	24	13	39	84	252	24- 0	7- 3	24- 0
38- 0	33- 5	23- 6	767	42	126	11	33	13	39	90	270	24- 0	10- 0	24- 0

NOTE: Larger sizes available for special designs.

FIGURE 7-10 Long-span high-profile arch pipe sizes and layout details.
(Courtesy: American Iron and Steel Institute.)

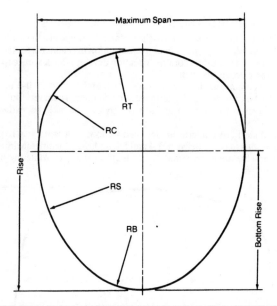

Max. Span, ft-in.	Rise, ft-in.	Rise Bottom, ft-in.	Area	Periphery										Inside Radius			
				Top		Corner		Side		Bottom		Total		Bottom Radius, ft-in.	Side Radius, ft-in.	Corner Radius, ft-in.	Top Radius, ft-in.
				N	Pi	N	Pi	N	Pi	N	Pi	N	Pi				
23- 8	25- 8	14-11	481	25	75	5	15	24	72	15	30	98	294	8-11	16- 7	6- 3	14- 8
24- 0	25-10	15- 1	496	22	66	7	21	22	66	20	60	100	300	9-11	17- 4	7- 0	16- 2
25- 6	25-11	15-10	521	27	81	7	21	20	60	21	63	102	306	10- 7	18- 1	6-11	15-10
24-10	27- 8	16- 9	544	27	81	5	15	25	75	18	54	105	315	9- 3	19- 8	5- 9	15-11
27- 5	27- 0	18- 1	578	30	90	6	18	26	78	16	48	110	330	9- 7	20- 4	4- 7	19-11
26- 8	28- 3	18- 0	593	28	84	5	15	30	90	12	36	110	330	8- 0	20- 1	4- 9	20-11
28- 1	27-10	16-10	624	27	81	8	24	22	66	25	75	112	336	12- 2	19- 0	7- 3	20- 5
28- 7	30- 7	19- 7	689	32	96	7	21	24	72	24	72	118	354	11- 2	24- 0	7- 0	18- 2
30- 0	29- 8	20- 0	699	32	96	8	24	23	69	25	75	119	357	11-11	24- 0	6- 7	21-10
30- 0	31- 2	19-11	736	34	102	7	21	24	72	26	78	122	366	12- 1	24- 0	7- 0	19- 3

FIGURE 7-11   Long-span pear-shape pipe sizes and layout details. *(Courtesy: American Iron and Steel Institute.)*

CORRUGATED STEEL BOX CULVERTS

Corrugated steel box culverts approach the rectangular shape of a low wide box. This is made possible by the addition of special reinforcing elements to standard structural plates. The resulting combined section develops the flexural capacity required by the extreme geometry.

Box culverts are available in spans up to 21 ft and rises to 10 ft-3 in. The foundation for box culverts can be designed as conventional concrete footers or, as partial steel spread footers, or as full steel invert. The table below lists all available sizes.

Corrugated steel box culverts are designed for low, wide waterway requirements with heights of cover between 1.4 and 5 ft and H-20 loadings. Minimum cover requirements are approximately one-tenth of the span while the maximum cover is 5 ft.

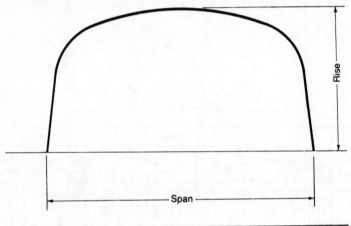

Rise, ft-in.	Span, ft-in.	Area ft^2	Rise, ft-in.	Span, ft-in.	Area ft^2
2-7	9-8	20.8	3-9	12-10	41.0
2-8	10-5	23.2	3-10	13-6	44.5
2-9	11-1	25.7	3-10	17-4	55.0
2-10	11-10	28.3	3-11	14-2	48.2
2-11	12-6	31.1	3-11	18-0	59.1
3-1	13-3	34.0	4-1	14-10	52.0
3-2	13-11	37.1	4-1	18-8	63.4
3-3	14-7	40.4	4-2	10-7	36.4
3-4	10-1	28.4	4-2	15-6	55.9
3-5	10-10	31.4	4-3	11-2	39.9
3-5	15-3	43.8	4-3	19-4	67.9
3-6	11-6	34.5	4-4	11-10	43.5
3-6	16-0	47.3	4-4	16-2	60.1
3-8	12-2	37.7	4-5	12-6	47.3
3-8	16-8	51.1	4-6	13-2	51.2

FIGURE 7-12   Layout details for corrugated steel box culverts — sizes and layout details. *(Courtesy: American Iron and Steel Institute.)*

Rise, ft-in.	Span, ft-in.	Area ft²	Rise, ft-in.	Span, ft-in.	Area ft²
4-6	16-10	64.4	6-9	13-7	77.9
4-7	17-6	68.9	6-9	16-9	99.3
4-7	20-8	77.6	6-10	14-2	83.3
4-8	13-10	55.3	6-10	17-4	105.1
4-9	14-6	59.5	7-0	14-9	88.9
4-9	18-1	73.5	7-0	17-11	111.1
4-10	15-1	63.8	7-0	20-8	127.2
4-11	11-0	44.7	7-1	15-4	94.6
4-11	18-9	78.4	7-2	18-6	117.3
5-0	11-7	48.7	7-3	12-3	71.5
5-0	15-9	68.3	7-3	15-10	100.5
5-1	12-3	52.9	7-4	12-10	77.1
5-1	16-4	73.0	7-4	16-5	106.5
5-1	19-5	83.4	7-4	19-1	123.6
5-2	12-10	57.2	7-5	13-5	82.8
5-3	17-0	77.8	7-6	13-11	88.6
5-4	13-6	61.7	7-6	17-0	112.7
5-5	14-1	66.2	7-8	14-6	94.5
5-5	17-7	82.8	7-8	17-6	119.0
5-5	20-8	94.1	7-9	15-0	100.6
5-6	14-9	71.0	7-9	18-1	125.5
5-7	18-3	88.0	7-11	15-7	106.8
5-8	11-5	53.3	7-11	18-7	132.1
5-8	15-4	75.8	8-0	12-8	81.1
5-8	18-10	93.4	8-0	16-1	113.1
5-9	12-0	57.9	8-1	19-2	138.9
5-9	16-0	80.9	8-2	16-8	119.6
5-10	12-7	62.6	8-2	13-9	93.3
5-10	19-6	98.9	8-3	19-8	145.9
5-11	16-7	86.1	8-4	17-2	126.2
6-0	13-3	67.4	8-5	14-10	106.0
6-1	13-10	72.4	8-5	17-8	133.0
6-1	17-2	91.4	8-7	18-3	139.9
6-2	14-5	77.5	8-7	20-9	160.3
6-2	17-9	96.9	8-8	15-10	119.2
6-2	20-8	110.6	8-9	18-9	147.0
6-4	15-0	82.7	8-11	16-10	132.9
6-4	18-4	102.6	8-11	19-3	154.2
6-5	11-10	62.2	9-1	19-9	161.6
6-5	15-7	88.1	9-3	17-10	147.1
6-6	18-11	108.5	9-5	20-9	176.9
6-7	12-5	67.3	9-6	18-10	162.0
6-7	16-2	93.6	9-10	19-10	177.4
6-8	13-0	72.5	10-2	20-9	193.5
6-8	19-6	114.5			

FIGURE 7-12 *(Continued)*

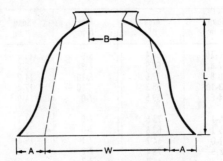

PLAN

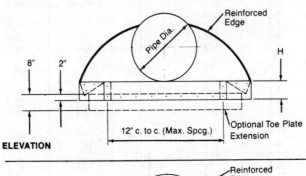

ELEVATION

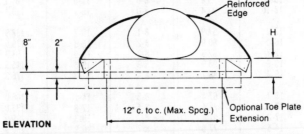

ELEVATION

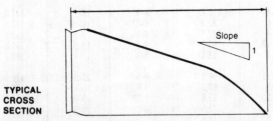

TYPICAL CROSS SECTION

FIGURE 7-13 Details of end sections for 2⅔-inch × ½-inch, 3-inch × 1-inch, and 5-inch × 1-inch round and pipe arch shapes. *(Courtesy: American Iron and Steel Institute.)*

TABLE 7-1 Dimensions of Galvanized Steel End Sections for (a) Round Pipe (2⅔-inch × ½-inch, 3-inch × 1-inch, and 5-inch × 1-inch Corrugations) and (b) Pipe Arch (2⅔-inch × ½-inch Corrugations) Shapes

Pipe Diameter, in.	Metal Thickness, in.	Dimensions, in.					Approximate Slope	Body
		A ± 1 in.	B (max.)	H ± 1 in.	L ± 1½ in.	W ± 2 in.		
12	0.064	6	6	6	21	24	2½	1 Pc.
15	0.064	7	8	6	26	30	2½	1 Pc.
18	0.064	8	10	6	31	36	2½	1 Pc.
21	0.064	9	12	6	36	42	2½	1 Pc.
24	0.064	10	13	6	41	48	2½	1 Pc.
30	0.079	12	16	8	51	60	2½	1 Pc.
36	0.079	14	19	9	50	72	2½	2 Pc.
42	0.109	16	22	11	69	84	2½	2 Pc.
48	0.109	18	27	12	78	90	2¼	2 Pc.
54	0.109	18	30	12	84	102	2	2 Pc.
60	0.109	18	33	12	87	114	1¾	3 Pc.
66	0.109	18	36	12	87	120	1½	3 Pc.
72	0.109	18	39	12	87	126	1⅓	3 Pc.
78	0.109	18	42	12	87	132	1¼	3 Pc.
84	0.109	18	45	12	87	138	1⅙	3 Pc.

(a)

Pipe-Arch, in.		Metal Thickness, in.	Dimensions, in.					Approximate Slope	Body
Span	Rise		A ± 1 in.	B (max.)	H ± 1 in.	L ± 1½ in.	W ± 2 in.		
17	13	0.064	7	9	6	19	30	2½	1 Pc.
21	15	0.064	7	10	6	23	36	2½	1 Pc.
24	18	0.064	8	12	6	28	42	2½	1 Pc.
28	20	0.064	9	14	6	32	48	2½	1 Pc.
35	24	0.079	10	16	6	39	60	2½	1 Pc.
42	29	0.079	12	18	8	46	75	2½	1 Pc.
49	33	0.109	13	21	9	53	85	2½	2 Pc.
57	38	0.109	18	26	12	63	90	2½	2 Pc.
64	43	0.109	18	30	12	70	102	2¼	2 Pc.
71	47	0.109	18	33	12	77	114	2¼	3 Pc.
77	52	0.109	18	36	12	77	126	2	3 Pc.
83	57	0.109	18	39	12	77	138	2	3 Pc.

(b)

NOTES:

1. All 3-piece bodies to have 0.109-in. sides and 0.138-in. center panels. Multiple panel bodies to have lap seams which are to be tightly joined by galvanized rivets or bolts.
2. For 60 thru 84 in. sizes, reinforced edges to be supplemented with galvanized stiffener angles. The angles to be attached by galvanized nuts and bolts.
   For the 77 by 52 in. and 83 by 57 in. sizes, reinforced edge to be supplemented by galvanized angles. The angles to be attached by galvanized bolts and nuts.
3. Angle reinforcement will be placed under the center panel seams on the 77 by 52 in. and 83 by 57 in. sizes.
4. Galvanized toe plate to be available as an accessory, when specified on the order and will be the same thickness as the End Section.

SOURCE: American Iron and Steel Institute.

7-1   Culvert Location Factors

Principles of culvert location

Culvert location means alignment and grade with respect to both roadway and stream. Proper location is important because it influences adequacy of the opening, maintenance of the culvert, and possible washout of the roadway. Although every culvert installation is a separate problem, the few principles set forth here apply in most cases.

A culvert is an enclosed channel serving as a continuation of and a substitute for an open stream where that stream meets an artificial barrier such as a roadway, embankment, or levee. It is necessary to consider abutting property, both as to ponding upstream and as to safe exit velocities, in order to avoid undue scour or silting downstream.

An open stream is not always stable. It may be changing its channel — straightening itself in some places, and becoming more sinuous in others. It may be scouring itself deeper in some places, silting in others. Change of land use upstream by clearing, deforestation, or real estate development can change both stability and flood flow of a stream.

Because a culvert is a *fixed* line in a stream, engineering judgment is necessary in properly locating the structure.

Alignment

The *first* principle of culvert location is to provide the stream with a direct entrance and a direct exit. Any abrupt change in direction at either end will retard the flow and make a larger structure necessary.

A direct inlet and outlet, if not already existing, can be obtained in one of three ways — by means of a channel change, a skewed alignment, or both. The cost of a channel change may be partly offset by a saving in culvert length or decrease in size. A skewed alignment requires a greater length of culvert, but is usually justified by improving the hydraulic condition and the safety of the roadbed.

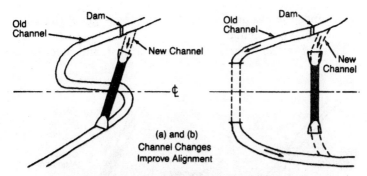

A channel change can provide more direct flow.

The *second* principle of culvert location is to use reasonable precaution to prevent the stream from changing its course near the ends of the culvert. Otherwise the culvert may become inadequate, cause excessive ponding, and possibly wash out — any one of which can lead to expensive maintenance of the roadway. Steel end sections, riprap, sod, or paving will help protect the banks from eroding and changing the channel.

Culvert alignment may also be influenced by choice of a grade line. Methods of selecting proper alignment are illustrated below.

At roadway intersections and private entrances, culverts should be placed in the direct line of the roadway ditch, especially where ditches are required to carry any considerable amount of storm water.

Culverts for drainage of cut-and-fill sections on long descending grades should be placed on a skew of about 45 degrees across the roadway. Thus the flow of water will not be retarded at the inlet.

Broken alignment under a roadway may be advisable on long culverts. Consideration should be given to entrance and exit conditions, and to increasing the size of the structure to handle or to cleaning out debris the stream may carry during flood periods.

Changes in alignment may be accomplished by welded miter cuts for bends, either in profile or alignment. The designer determines the radius of curvature and the angle of the miter cut can then be calculated.

This subsection courtesy of American Iron and Steel Institute.

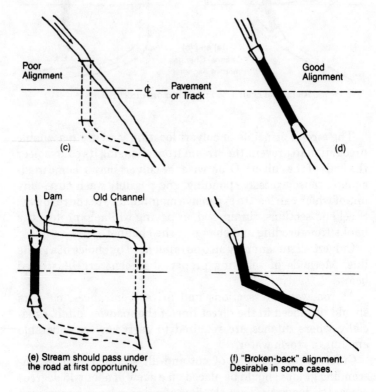

Various methods of obtaining correct culvert alignment.

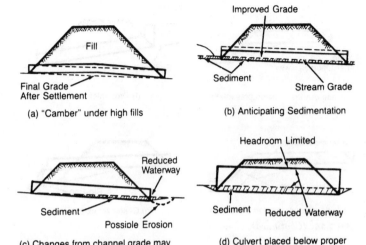

(a) "Camber" under high fills

(b) Anticipating Sedimentation

(c) Changes from channel grade may cause sedimentation or erosion

(d) Culvert placed below proper grade; waterway is reduced

FIGURE 7-14 Proper culvert grades. The ideal grade line for a culvert is one that produces neither silting nor excessive velocities and scour, one that gives the shortest length, and one that makes replacement simplest. Velocities as great as 10 feet per second cause destructive scour downstream and to the culvert structure itself unless protected. Safe streambed velocities are given in Table 7-13 on page 232. The silt carrying capacity of a stream varies as the square of the velocity. Capacity of a culvert with a free outlet (not submerged) is not increased by placing on a slope steeper than its critical slope (about 1 percent for a 96-inch pipe). The capacity is controlled by the amount of water that can get through the inlet.

On the other hand, the capacity of a pipe on a very slight gradient and with a *submerged* outlet is influenced by the head (difference in elevation of water surface at both ends). In this case, the roughness of the culvert interior, in addition to the velocity head and entrance loss, is a factor.

A slope of 1 to 2 percent is advisable to give a gradient equal to or greater than the critical slope, provided the velocity falls within permissible limits. In general, a minimum slope of 0.5 foot in 100 feet will avoid sedimentation. In ordinary practice the grade line coincides with the average streambed above and below the culvert. However, deviation for a good purpose is permissible.

(Courtesy: American Iron and Steel Institute.)

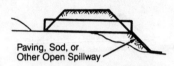

Paving, Sod, or
Other Open Spillway

(e) Hillside grades; erosion prevention

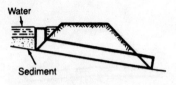

Water

Sediment

(f) Drop Inlet

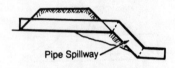

Pipe Spillway

(g) Hillside grades; erosion prevention

FIGURE 7-14 *(Continued)*

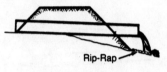

Rip-Rap

(h) Cantilever extension

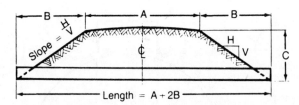

Computation of culvert length when flow line is on a flat grade.

(a)

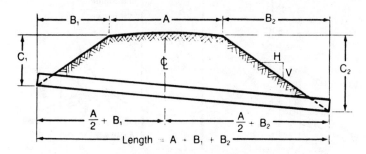

Determining culvert length on a steep grade.

(b)

FIGURE 7-15 Culvert length. The required length of a culvert depends on the width of the roadway or roadbed, the height of fill, the slope of the embankment, the slope and skew of the culvert, and the type of end finish such as end section, headwall, beveled end, drop inlet, or spillway. A culvert should be long enough so that its ends do not clog with sediment or become covered with a settling, spreading embankment.

A cross-sectional sketch of the embankment and a profile of the streambed is the best way to determine the length of culvert needed. Lacking such a sketch, the length of a simple culvert under an embankment can be determined as follows: To the width of the roadway (and shoulders), add twice the slope ratio times the height of fill at the center of the road. The height of fill should be measured to the flow line if headwalls are not to be used, and to the top of the culvert if headwalls or end sections are to be installed.

Example: A roadway is 40 feet wide on top, two to one side slopes, and at the center of the road the height of fill to flow line is 7 feet: $40 + (4 \times 7) = 68$ feet length at flow line (*a*).

If the culvert is on a slope of 5 percent or more, it may be advisable to compute the sloped length in the manner shown in (*b*). However, as fill slopes usually vary from the established grade stakes, any refinement in computing culvert length may not be necessary.

(Courtesy: American Iron and Steel Institute.)

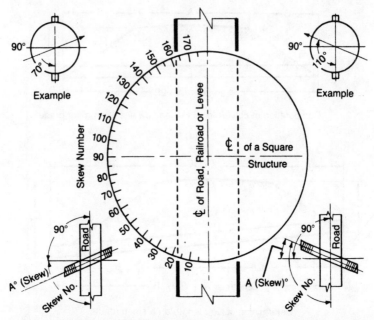

Diagram for method of properly specifying skewed culverts. The direction of flow should also be indicated for fabrication as a left or right.

FIGURE 7-16   Pipe length for skewed culverts. Where a culvert crosses the roadway at other than a right angle, the increased bottom center line length should be computed as follows: First determine the length at right angles to the roadway. Then divide by the cosine of the angle between the normal and skewed direction. Correction for pipe diameter on a skew may be made by multiplying diameter by tangent of angle between normal and skew. If the roadway is on a horizontal or vertical curve or on a steep gradient, the additional length may be estimated.

Example: Assume a normal length of 62 feet and an angle of 70 degrees skewed from the normal: skew length = 62/0.342 = 181 feet.

The bottom center line length is specified to the nearest 2-foot length. The ends of the structure may be cut to make them parallel to the center line of the road. For correct fabrication of corrugated steel culverts it is essential to specify the *direction of flow* as well as the *skew angle* or skew number.

(Courtesy: American Iron and Steel Institute.)

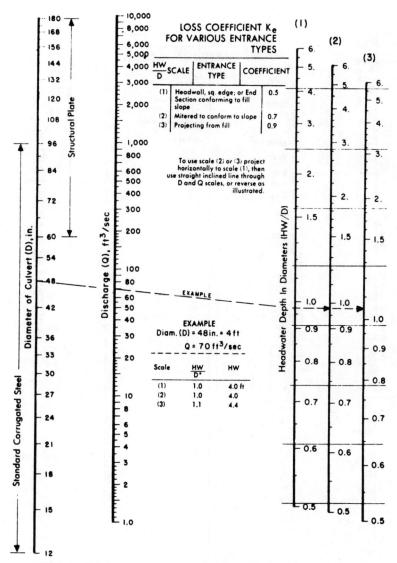

FIGURE 7-17 Headwater depth for corrugated steel culverts with inlet control. The manufacturers recommended keeping HW/D to a maximum of 1.5 and preferably to no more than 1.0. *(Courtesy: American Iron and Steel Institute.)*

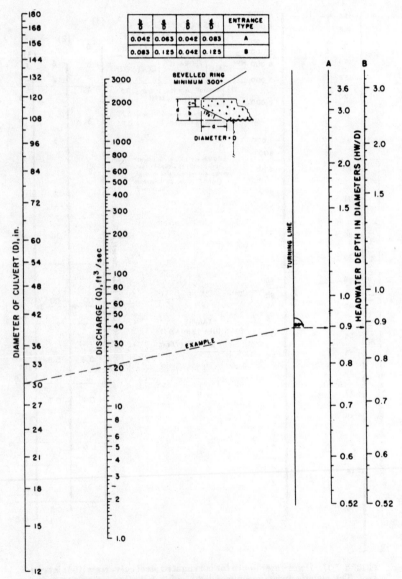

FIGURE 7-18 Headwater depth for circular culverts with beveled ring inlet control. *(Courtesy: American Iron and Steel Institute.)*

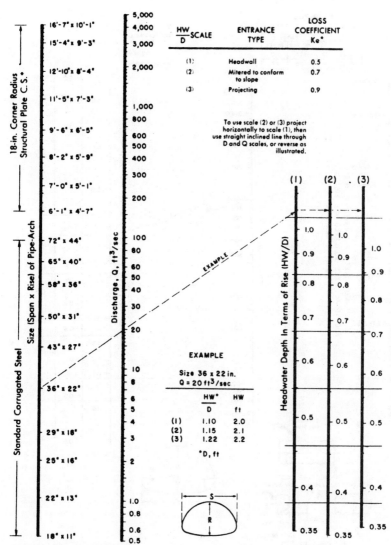

$\dfrac{HW}{D}$ SCALE	ENTRANCE TYPE	LOSS COEFFICIENT K_e*
(1)	Headwall	0.5
(2)	Mitered to conform to slope	0.7
(3)	Projecting	0.9

To use scale (2) or (3) project horizontally to scale (1), then use straight inclined line through D and Q scales, or reverse as illustrated.

EXAMPLE

Size 36 x 22 in.
Q = 20 ft³/sec

	$\dfrac{HW^*}{D}$	HW ft
(1)	1.10	2.0
(2)	1.15	2.1
(3)	1.22	2.2

*D, ft

*Additional Sizes Not Dimensioned Are Listed In Fabricator's Catalog

FIGURE 7-19 Headwater depth for corrugated steel pipe arch culverts with inlet control. Headwater depth should be kept low because pipe arches are generally used where headroom is limited. *(Courtesy: American Iron and Steel Institute.)*

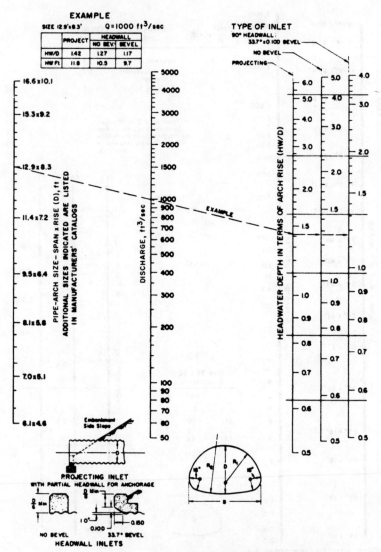

FIGURE 7-20   Headwater depth for inlet control structural plate pipe arch culverts. *(Courtesy: American Iron and Steel Institute.)*

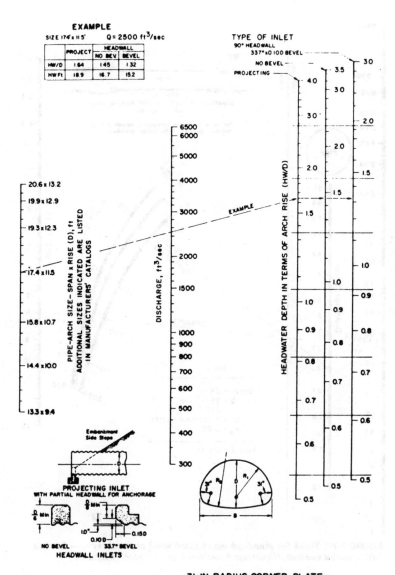

31-IN. RADIUS CORNER PLATE.
PROJECTING OR HEADWALL INLET
HEADWALL WITH OR WITHOUT EDGE BEVEL

FIGURE 7-20 *(Continued)*

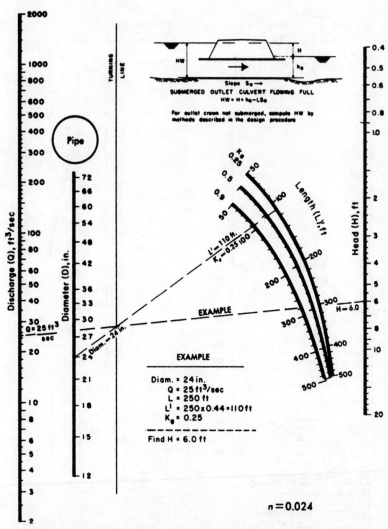

FIGURE 7-21 Head for standard corrugated steel pipe culverts—flowing full—outlet control. *(Courtesy: American Iron and Steel Institute.)*

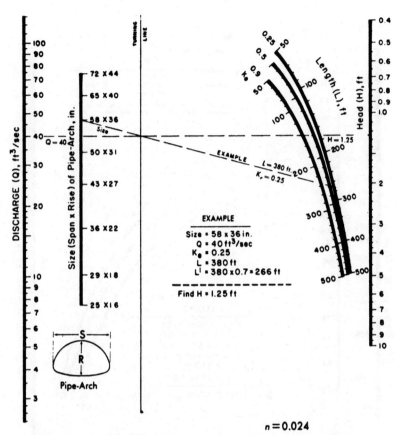

$n = 0.024$

FIGURE 7-22 Head for standard corrugated steel pipe arch culverts—flowing full—outlet control. *(Courtesy: American Iron and Steel Institute.)*

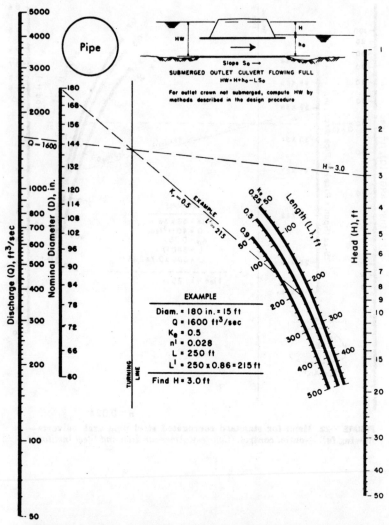

FIGURE 7-23 Head for structural plate corrugated steel pipe culverts—flowing full—outlet control. *(Courtesy: American Iron and Steel Institute.)*

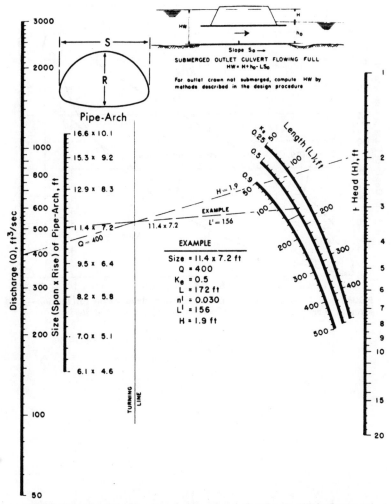

FIGURE 7-24 Head for structural plate pipe arch culverts, 18-inch corner radius — flowing full — outlet control. For 31-inch corner radius, use structure sizes with equivalent areas on the 18-inch corner radius scale. *(Courtesy: American Iron and Steel Institute.)*

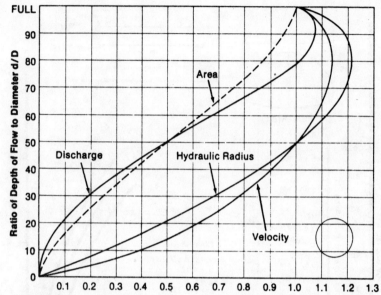

FIGURE 7-25 Hydraulic elements in terms of hydraulics for full section—circular corrugated steel pipe. *(Courtesy: American Iron and Steel Institute.)*

TABLE 7-2 Full-Flow Data for Round Pipe

Diameter, in.	Area, ft²	Hydraulic Radius, ft	Diameter, in.	Area, ft²	Hydraulic Radius, ft
12	0.785	0.250	156	132.7	3.25
15	1.227	0.3125	162	143.1	3.375
18	1.767	0.375	168	153.9	3.5
21	2.405	0.437	174	165.1	3.625
24	3.142	0.50	180	176.7	3.75
30	4.909	0.625	186	188.7	3.875
36	7.069	0.75	192	201.1	4.0
42	9.621	0.875	198	213.8	4.125
48	12.566	1.0	204	227.0	4.25
54	15.904	1.125	210	240.5	4.375
60	19.635	1.25	216	254.5	4.5
66	23.758	1.375	222	268.8	4.625
72	28.27	1.5	228	283.5	4.75
78	33.18	1.625	234	298.6	4.875
84	38.49	1.75	240	314.2	5.0
90	44.18	1.875	246	330.1	5.125
96	50.27	2.0	252	346.4	5.25
108	63.62	2.25	258	363.1	5.375
114	70.88	2.375	264	380.1	5.5
120	78.54	2.5	270	397.6	5.625
126	86.59	2.625	276	415.5	5.75
132	95.03	2.75	282	433.7	5.875
138	103.87	2.875	288	452.4	6.0
144	113.10	3.00	294	471.4	6.125
150	122.7	3.125	300	490.9	6.25

SOURCE: American Iron and Steel Institute.

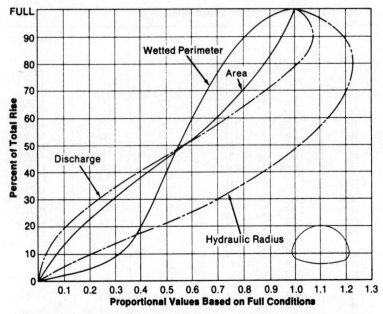

FIGURE 7-26 Hydraulic properties of corrugated steel and structural plate pipe arches. *(Courtesy: American Iron and Steel Institute.)*

TABLE 7-3   Full-Flow Data for Corrugated Steel Pipe Arches

	Corrugations 2⅔ × ½ in.				Corrugations 3 × 1 in. and 5 × 1 in.			
Dimensions, in.								
Pipe Diameter	Pipe-Arch		Waterway Area, ft²	Hydraulic Radius A/πD, ft	Diameter, in.	Equiv. Size, in.	Waterway Area, ft²	Hydraulic Radius A/πD, ft
	Span	Rise						
15	17	13	1.1	0.280	54	60 × 46	15.6	1.104
18	21	15	1.6	0.340	60	66 × 51	19.3	1.230
21	24	18	2.2	0.400	66	73 × 55	23.2	1.343
24	28	20	2.9	0.462	72	81 × 59	27.4	1.454
30	35	24	4.5	0.573	78	87 × 63	32.1	1.573
36	42	29	6.5	0.69	84	95 × 67	37.0	1.683
*42	49	33	8.9	0.81	90	103 × 71	42.4	1.80
*48	57	38	11.6	0.924	96	112 × 75	48.0	1.911
*54	64	43	14.7	1.04	102	117 × 79	54.2	2.031
*60	71	47	18.1	1.153	108	128 × 83	60.5	2.141
*66	77	52	21.9	1.268	114	137 × 87	67.4	2.259
*72	83	57	26.0	1.38	120	142 × 91	74.5	2.373

SOURCE: American Iron and Steel Institute.

TABLE 7-4   Full-Flow Data for Structural Plate Pipe Arches
[Corrugations, 6 × 2 inches; Corner Plates, 9 pi; Radius (R_c), 18 inches]

Dimensions, ft-in.		Waterway Area, ft²	Hydraulic Radius, ft
Span	Rise		
6-1	4-7	22	1.29
6-4	4-9	24	1.35
6-9	4-11	26	1.39
7-0	5-1	28	1.45
7-3	5-3	30	1.51
7-8	5-5	33	1.55
7-11	5-7	35	1.61
8-2	5-9	38	1.67
8-7	5-11	40	1.71
8-10	6-1	43	1.77
9-4	6-3	45	1.81
9-6	6-5	48	1.87
9-9	6-7	51	1.93
10-3	6-9	54	1.97
10-8	6-11	57	2.01
10-11	7-1	60	2.07
11-5	7-3	63	2.11
11-7	7-5	66	2.17
11-10	7-7	70	2.23
12-4	7-9	73	2.26
12-6	7-11	77	2.32
12-8	8-1	81	2.38
12-10	8-4	85	2.44
13-5	8-5	88	2.48
13-11	8-7	91	2.52
14-1	8-9	95	2.57
14-3	8-11	100	2.63
14-10	9-1	103	2.67
15-4	9-3	107	2.71
15-6	9-5	111	2.77
15-8	9-7	116	2.83
15-10	9-10	121	2.89
16-5	9-11	125	2.92
16-7	10-1	130	2.98

SOURCE: American Iron and Steel Institute.

TABLE 7-5 Full-Flow Data for Corrugated Steel Pipe Arches
[Corrugations, 6 × 2 inches; Corner Plates, 15 pi; Radius (R_c), 31 inches]

Span, ft-in.	Rise, ft-in.	Area, ft²	Hydraulic Radius, ft
13-3	9-4	97	2.68
13-6	9-6	102	2.74
14-0	9-8	105	2.78
14-2	9-10	109	2.83
14-5	10-0	114	2.90
14-11	10-2	118	2.94
15-4	10-4	123	2.98
15-7	10-6	127	3.04
15-10	10-8	132	3.10
16-3	10-10	137	3.14
16-6	11-0	142	3.20
17-0	11-2	146	3.24
17-2	11-4	151	3.30
17-5	11-6	157	3.36
17-11	11-8	161	3.40
18-1	11-10	167	3.45
18-7	12-0	172	3.50
18-9	12-2	177	3.56
19-3	12-4	182	3.59
19-6	12-6	188	3.65
19-8	12-8	194	3.71
19-11	12-10	200	3.77
20-5	13-0	205	3.81
20-7	13-2	211	3.87

SOURCE: American Iron and Steel Institute.

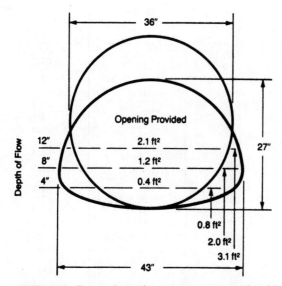

FIGURE 7-27 Comparison of waterway cross-sectional areas at equal depths of flow in steel pipe and pipe arch. The pipe arch handles a larger volume at the lower levels of flow. *(Courtesy: American Iron and Steel Institute.)*

TABLE 7-6 Full-Flow Data for Structural Plate Arches

Dimension[1]		Waterway Area, ft²	Hydraulic Radius, ft
Span, ft	Rise, ft-in.		
6.0	1-9½	7½	0.575
	2-3½	10	0.722
	3-2	15	0.973
7.0	2-4	12	0.767
	2-10	15	0.914
	3-8	20	1.111
8.0	2-11	17	0.934
	3-4	20	1.053
	4-2	26	1.264
9.0	2-11	18½	0.925
	3-10½	26½	1.229
	4-8½	33	1.427
10.0	3-5½	25	1.108
	4-5	34	1.409
	5-3	41	1.595
11.0	3-6	27½	1.219
	4-5½	37	1.427
	5-9	50	1.769
12.0	4-0½	35	1.300
	5-0	45	1.579
	6-3	59	1.913
13.0	4-1	38	1.324
	5-1	49	1.618
	6-9	70	2.095
14.0	4-7½	47	1.503
	5-7	58	1.766
	7-3	80	2.223
15.0	4-7½	50	1.513
	5-8	62	1.791
	6-7	75	2.072
	7-9	92	2.386
16.0	5-2	60	1.684
	7-1	86	2.218
	8-3	105	2.554
17.0	5-2½	63	1.684
	7-2	92	2.184
	8-10	119	2.724
18.0	5-9	75	1.876
	7-8	104	2.412
	8-11	126	2.771
19.0	6-4	87	2.045
	8-2	118	2.583
	9-5½	140	2.914
20.0	6-4	91	2.053
	8-3½	124	2.612
	10-0	157	3.102
21.0	6-11	104	2.217
	8-10	140	2.797
	10-6	172	3.234

[1]Dimensions are to inside crests and are subject to manufacturing tolerances.

TABLE 7-6 **Full-Flow Data for Structural Plate Arches** *(Continued)*

Dimension[1]		Waterway Area, ft²	Hydraulic Radius, ft
Span, ft	Rise, ft-in.		
22.0	6-11	109	2.239
	8-11	146	2.817
	11-0	190	3.408
23.0	8-0	134	2.575
	9-10	171	3.099
	11-6	208	3.566
24.0	8-6	150	2.746
	10-4	188	3.255
	12-0	226	3.711
25.0	8-6½	155	2.748
	10-10½	207	3.431
	12-6	247	3.892

[1]Dimensions are to inside crests and are subject to manufacturing tolerances.

SOURCE: American Iron and Steel Institute.

TABLE 7-7   Hydraulic Data for Long-Span Horizontal Ellipse

Span × Rise (B × D) ft-in.	Full Flow Data				Discharge — (Q), ft³/sec					
	Area, ft²	WP, ft	R, ft	AR²/³	Critical Depth					
					0.4D	0.5D	0.6D	0.7D	0.8D	0.9D
19-4 × 12-9	191	50.7	3.77	462.7	769	1204	1714	2316	3083	4183
20-1 × 13-0	202	52.3	3.86	497.1	823	1282	1832	2478	3298	4502
20-2 × 11-11	183	50.7	3.61	430.6	708	1110	1584	2153	2871	3935
20-10 × 12-2	194	52.3	3.71	464.9	756	1154	1694	2298	3088	4194
21-0 × 15-2	248	57.1	4.35	660.9	1073	1684	2390	3225	4336	5901
21-11 × 13-1	221	55.5	3.98	555.0	897	1403	2005	2725	3640	4966
22-6 × 15-8	274	60.3	4.55	752.4	1228	1921	2732	3687	4886	5979
23-0 × 14-1	249	58.7	4.25	653.3	1051	1645	2347	3185	4256	5801
23-3 × 15-11	288	61.9	4.65	802.3	1298	2033	2889	3903	5179	7046
24-4 × 16-11	320	65.1	4.92	925.7	1486	2327	3307	4464	5914	8055
24-6 × 14-8	274	61.9	4.43	739.1	1177	1843	2634	3577	4785	6518
25-2 × 14-11	287	63.5	4.53	785.7	1242	1947	2782	3780	5060	6881
25-5 × 16-9	330	66.7	4.95	958.5	1523	2383	3391	4588	6101	8295
26-1 × 18-2	369	69.9	5.28	1118.9	1775	2778	3949	5331	7046	9611
26-3 × 15-10	320	66.9	4.80	910.6	1430	2240	3196	4340	5804	7902
27-0 × 16-2	334	68.3	4.89	962.2	1503	2356	3366	4572	6113	8303
27-2 × 19-1	405	73.1	5.54	1268.0	1999	3131	4448	6004	7953	10817
27-11 × 19-5	421	74.7	5.64	1334.0	2095	3278	4660	6290	8329	11325
28-1 × 17-1	369	71.5	5.16	1101.9	1714	2683	3830	5192	6943	9438
28-10 × 17-5	384	73.1	5.26	1161.4	1795	2812	4016	5452	7288	9919
29-5 × 19-11	455	77.9	5.84	1475.5	2289	3587	5098	6887	9143	12434
30-1 × 20-2	472	79.5	5.94	1548.1	2391	3744	5326	7198	9563	13008
30-3 × 17-11	415	76.3	5.44	1283.6	1968	3084	4406	5985	8014	10900
31-2 × 21-2	513	82.7	6.20	1731.3	2659	4166	5925	8003	10622	14429
31-4 × 18-11	454	79.5	5.71	1450.4	2212	3467	4950	6720	8983	12205
32-1 × 19-2	471	81.1	5.81	1522.2	2309	3617	5173	7020	9389	12774
32-3 × 22-2	555	85.9	6.46	1925.1	2947	4615	6561	8870	11752	15964
33-0 × 22-5	574	87.5	6.56	2011.5	3064	4798	6825	9220	12236	16639
33-2 × 20-1	512	84.3	6.08	1705.6	2577	4038	5762	7819	10451	14210
34-1 × 23-4	619	90.7	6.82	2226.1	3376	5286	7495	10150	13464	18280
34-7 × 20-8	548	87.5	6.26	1861.4	2792	4372	6245	8481	11320	15424
34-11 × 21-4	574	89.1	6.44	1986.9	2975	4661	6652	9070	12053	16397
35-1 × 24-4	665	93.9	7.08	2452.0	3705	5801	8246	11131	14754	20048
36-0 × 22-4	619	92.3	6.71	2202.1	3286	5146	7341	9950	13283	18054
37-2 × 22-2	631	93.9	6.72	2247.0	3328	5215	7450	10100	13537	18381

SOURCE: American Iron and Steel Institute.

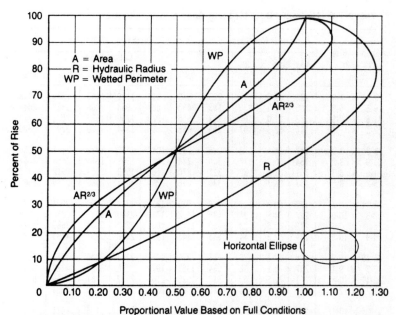

FIGURE 7-28 Hydraulic properties of long-span horizontal ellipse. *(Courtesy: American Iron and Steel Institute.)*

TABLE 7-8 Hydraulic Data for Long-Span Low-Profile Arch

Span × Rise (B × D) ft-in.	Full Flow Data				Discharge — (Q). ft³/sec					
	Area, ft²	WP, ft	R, ft	AR²/³	Critical Depth					
					0.4D	0.5D	0.6D	0.7D	0.8D	0.9D
20-1 × 7-6	120	47.9	2.51	223	579	819	1119	1443	1839	2448
19-5 × 6-9	105	45.6	2.30	183	480	681	933	1206	1532	2049
21-6 × 7-9	133	51.0	2.62	253	656	929	1266	1632	2081	2723
22-3 × 7-11	140	52.5	2.67	269	698	986	1344	1720	2207	2926
23-0 × 8-0	147	54.1	2.72	286	738	1042	1420	1829	2324	3083
23-9 × 8-2	154	55.6	2.77	304	784	1112	1508	1944	2466	3267
24-6 × 8-3	161	57.2	2.82	321	830	1180	1593	2054	2612	3422
25-2 × 8-5	168	58.7	2.86	339	876	1249	1682	2166	2756	3640
25-11 × 8-7	176	60.2	2.91	359	923	1320	1774	2290	2901	3848
27-3 × 10-0	217	64.8	3.34	485	1228	1733	2340	3017	3836	5046
28-1 × 9-6	212	65.6	3.23	463	1179	1666	2252	2901	3685	4858
28-9 × 10-3	234	67.9	3.44	533	1343	1896	2558	3297	4193	5538
28-10 × 9-8	220	67.1	3.28	486	1236	1747	2361	3040	3863	5105
30-3 × 9-11	237	70.2	3.38	534	1353	1914	2586	3326	4220	5543
30-11 × 10-8	261	72.5	3.59	612	1536	2168	2922	3761	4760	6292
31-7 × 12-1	309	76.1	4.06	786	1901	2684	3607	4676	5961	7836
31-0 × 10-1	246	71.7	3.43	560	1416	2004	2702	3476	4411	5806
32-4 × 12-3	319	77.6	4.11	819	1979	2795	3760	4869	6201	8142
31-9 × 10-2	255	73.3	3.47	585	1480	2099	2824	3631	4611	6086
33-1 × 12-5	330	79.1	4.17	855	2085	2907	3916	5065	6443	8500
33-2 × 11-1	289	77.1	3.74	696	1741	2459	3313	4264	5413	7099
34-5 × 13-3	367	83.0	4.42	988	2372	3346	4467	5815	7400	9730
34-7 × 11-4	308	80.2	3.84	755	1886	2674	3589	4614	5853	7722
37-11 × 15-7	477	92.9	5.13	1419	3356	4726	6290	8158	10390	13646
35-4 × 11-5	318	81.7	3.89	787	1962	2784	3730	4782	6089	8015
38-8 × 15-9	490	94.4	5.19	1469	3466	4881	6496	8426	10727	14122

SOURCE: American Iron and Steel Institute.

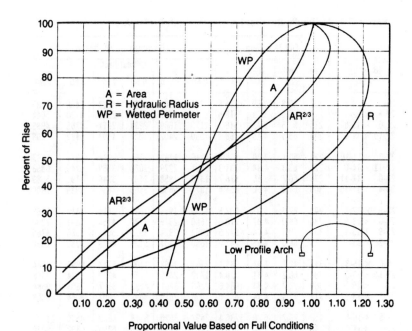

FIGURE 7-29 **Hydraulic properties of long-span low-profile arch.** *(Courtesy: American Iron and Steel Institute.)*

TABLE 7-9   Hydraulic Data for Long-Span High-Profile Arch

Span × Rise (B × D) ft-in.	Full Flow Data				Discharge — (Q), ft³/sec					
	Area, ft²	WP, ft	R, ft	AR²/³	Critical Depth					
					0.4D	0.5D	0.6D	0.7D	0.8D	0.9D
20-1 × 9-1	152	50.8	2.99	315.5	785	1107	1480	1923	2466	3282
20-8 × 12-1	214	56.5	3.78	518.6	1191	1687	2264	2936	3790	5044
21-6 × 11-8	215	57.5	3.73	516.1	1179	1669	2234	2911	3765	4989
22-10 × 14-6	284	63.9	4.45	768.9	1690	2402	3227	4193	5412	7209
22-3 × 11-10	224	59.1	3.80	546.5	1246	1762	2361	3077	3974	5297
22-11 × 14-0	275	63.3	4.34	731.3	1601	2279	3050	3969	5140	6853
23-0 × 11-11	234	60.7	3.86	576.7	1315	1858	2491	3246	4191	5589
24-4 × 14-10	309	67.2	4.60	854.4	1874	2658	3569	4636	5989	7980
23-9 × 12-1	244	62.3	3.93	608.8	1385	1956	2623	3418	4417	5869
24-6 × 13-9	288	66.0	4.37	770.5	1680	2376	3187	4154	5391	7200
25-9 × 15-1	334	70.5	4.74	942.5	2063	2924	3923	5096	6586	8769
25-2 × 13-1	283	66.6	4.25	742.0	1650	2331	3125	4079	5280	7030
26-6 × 15-3	347	72.1	4.81	988.1	2161	3062	4106	5312	6896	9184
25-11 × 13-3	294	68.2	4.31	778.4	1730	2445	3276	4280	5534	7348
27-3 × 15-5	360	73.7	4.88	1034.6	2260	3201	4292	5577	7803	9584
27-5 × 13-6	317	71.3	4.44	855.0	1896	2679	3591	4692	6064	8068
29-5 × 16-5	412	79.2	5.20	1235.4	2697	3820	5118	6639	8570	11390
28-2 × 14-5	348	74.0	4.70	976.0	2123	2998	4019	5255	6802	9050
30-1 × 18-0	466	82.8	5.63	1474.0	3111	4402	5920	7694	9952	13266
30-3 × 15-5	399	79.5	5.02	1169.0	2539	3589	4811	6278	8114	10775
31-7 × 18-4	496	86.1	5.77	1596.6	3366	4768	6405	8315	10760	14291
31-0 × 15-7	412	81.1	5.08	1216.8	2642	3734	5004	6534	8437	11173
31-8 × 17-9	483	85.4	5.65	1531.5	3222	4556	6114	7960	10323	13760
32-4 × 19-11	553	90.0	6.18	1863.2	3808	5404	7259	9450	12256	16350
31-9 × 17-2	469	84.8	5.53	1466.4	3080	4353	5836	7615	9890	13190
33-1 × 20-1	570	91.2	6.25	1934.9	3940	5610	7534	9807	12721	16963
32-6 × 17-4	484	86.4	5.60	1524.8	3200	4522	6061	7917	10270	13675
33-10 × 20-3	587	92.9	6.33	2009.9	4106	5820	7814	10172	13197	17607
34-0 × 17-8	513	89.6	5.73	1643.2	3445	4867	6524	8532	11054	14703
34-7 × 19-10	590	93.9	6.28	2007.6	4095	5797	7775	10136	13176	17575
34-8 × 17-9	528	91.2	5.79	1703.0	3572	5043	6762	8844	11458	15210
35-4 × 20-0	607	95.5	6.35	2080.4	4255	6022	8076	10534	13697	18270

SOURCE: American Iron and Steel Institute.

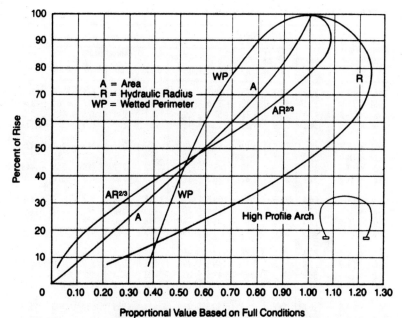

FIGURE 7-30 Hydraulic properties of long-span high-profile arch. *(Courtesy: American Iron and Steel Institute.)*

TABLE 7-10 Hydraulic Data for Structural Plate Box Culverts

Rise × Span (D × B) ft-in.	Full Flow Data				Discharge — (Q), ft³/sec					
	Area, ft²	WP, ft	R, ft	AR²/³	Critical Depth					
					0.4D	0.5D	0.6D	0.7D	0.8D	0.9D
2-7 × 9-8	20.8	21.9	0.95	20.1	58	81	106	134	166	207
2-11 × 12-6	31.1	27.9	1.12	33.5	93	129	169	213	264	334
3-5 × 15-3	43.8	33.8	1.29	51.9	138	193	254	323	405	540
3-6 × 11-6	34.5	26.9	1.28	40.7	111	154	202	255	316	393
3-11 × 14-2	48.2	32.7	1.47	62.3	163	226	297	376	467	598
3-11 × 18-0	59.1	39.7	1.49	77.0	207	288	379	483	619	824
4-2 × 10-7	36.4	25.9	1.40	45.6	127	172	227	288	359	448
4-6 × 13-2	51.2	31.7	1.61	70.5	178	251	330	419	520	650
4-6 × 16-10	64.4	38.6	1.67	90.6	223	315	418	533	670	898
4-7 × 20-8	77.6	45.6	1.70	110.7	291	409	542	701	903	1194
5-0 × 15-9	68.3	37.5	1.82	101.9	249	348	460	586	731	949
5-1 × 12-3	52.9	30.8	1.72	75.9	198	275	362	458	569	708
5-1 × 19-5	83.4	44.3	1.88	127.1	313	441	586	750	966	1291
5-6 × 14-9	71.0	36.5	1.94	110.6	270	383	502	636	790	992
5-7 × 18-3	88.0	43.2	2.04	141.5	335	473	628	802	1011	1359
6-1 × 17-2	91.4	42.1	2.17	153.2	367	505	668	852	1066	1399
6-2 × 20-8	110.6	48.8	2.27	190.9	459	644	853	1092	1406	1879
6-5 × 11-10	62.2	32.0	1.94	96.9	268	372	486	615	760	947
6-8 × 19-6	114.5	47.7	2.40	205.4	480	670	888	1135	1434	1931
6-10 × 14-2	83.3	37.5	2.22	141.8	356	495	648	825	1020	1270
7-4 × 16-5	106.5	43.0	2.48	195.1	462	644	854	1079	1341	1688
7-5 × 13-5	82.8	36.8	2.25	142.2	379	526	688	862	1063	1321
7-11 × 15-7	106.8	42.1	2.53	198.5	490	681	892	1116	1383	1727
7-11 × 18-7	132.1	48.4	2.73	258.2	590	824	1092	1388	1735	2274
8-0 × 12-8	81.1	36.1	2.25	139.3	398	552	720	901	1103	1371
8-5 × 14-10	106.0	41.4	2.56	198.4	508	706	923	1161	1429	1774
8-5 × 17-8	133.0	47.5	2.80	264.4	613	854	1120	1428	1776	2254
8-7 × 20-9	160.3	53.7	2.99	332.5	748	1046	1378	1761	2224	2996
8-11 × 16-10	132.9	46.6	2.85	267.1	634	883	1156	1470	1819	2276
9-1 × 19-9	161.6	52.7	3.07	341.2	773	1079	1406	1790	2245	2973
9-6 × 18-10	162.0	51.8	3.13	346.4	786	1096	1438	1834	2284	2936
10-2 × 20-9	193.5	56.9	3.40	437.5	965	1348	1779	2263	2842	3792

SOURCE: American Iron and Steel Institute.

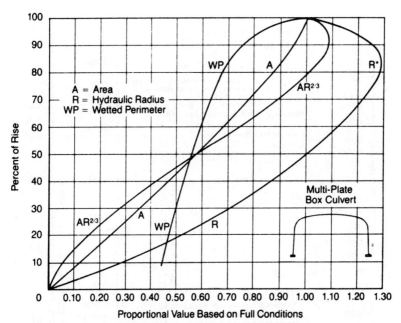

FIGURE 7-31 Hydraulic properties of structural plate box culverts. *(Courtesy: American Iron and Steel Institute.)*

Channel Description	Cross-Sectional Area of Flow	Wetted Perimeter	Best Hydraulic Depth
Rectangular Channel	$A = bh$	$\begin{aligned} p' &= 2h + b \\ &= 2h + \frac{A}{h} \end{aligned}$	$\begin{aligned} H_{\text{I}} &= \frac{bh}{2h + b} = \frac{2h^2}{4h} \\ &= h/2 \end{aligned}$
Trapezoidal Channel	$\begin{aligned} A &= h\,(zh + b') \\ &= h\,(zh + p' - \\ & \quad 2h\sqrt{(z^2 + 1)}) \end{aligned}$	$p' = 2h\sqrt{(z^2 + 1)} + b'$ where z is the side slope; if z is a variable and 'A' and 'h' are fixed, then — $p' = \frac{A}{h} - zh + 2h\sqrt{(z^2 + 1)}$	$H_{\text{I}} = h/2$; with $\theta = 60°$ for the best hydraulic section

FIGURE 7-32 Formulas for rectangular and trapezoidal channels. *(From Fluid Flow Pocket Handbook by Nicholas P. Cheremisinoff. Copyright 1984, Gulf Publishing Co., Houston, Texas. Used with permission. All rights reserved.)*

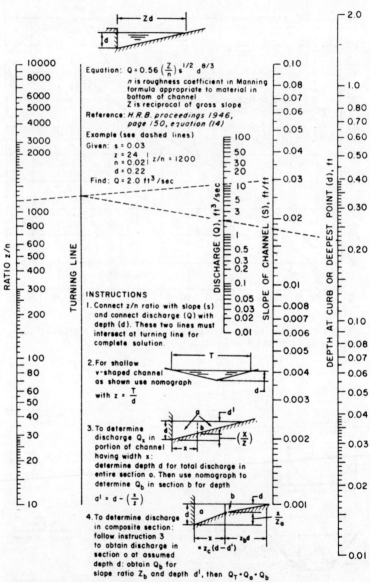

FIGURE 7-33   Nomograph for flow in triangular channels. *(Courtesy: American Iron and Steel Institute.)*

In large watersheds Tc should be obtained from stream flow records or field inspection of the stream channel and watershed characteristics. When such data is unavailable, the following empirical equation for concentration may be used for estimating the time of concentration:

Kirpich (1940)
$$Tc = 0.00013 \frac{L^{0.77}}{S^{0.385}}$$

in which Tc = time of concentration, hrs
L = maximum length of travel of water, ft
S = slope, equal to H/L where H is the difference in elevation between the most remote point on the basin and the outlet, ft/ft

Note: Use Kirpich equation for natural basins with well defined channels for overland flow on bare earth, and mowed grass roadside channels. For overland flow, grassed surfaces, multiply Tc by 2. For overland flow, concrete or ashphalt surfaces, multiply Tc by 0.4. For concrete channels, multiply Tc by 0.2.

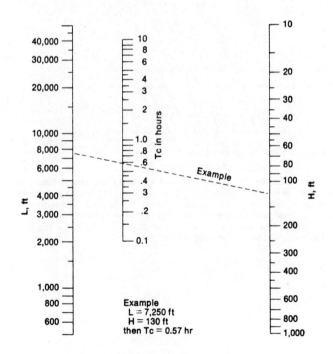

Tc nomograph for large watersheds.

FIGURE 7-34 Time of concentration for large watersheds. *(Courtesy: American Iron and Steel Institute.)*

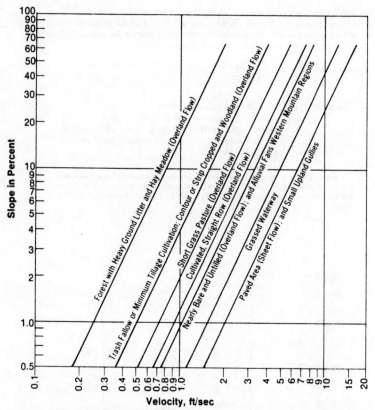

FIGURE 7-35 Velocities for upland method of estimating travel times for overland flow. When calculating travel times for overland flow in watersheds with a variety of land covers, the upland method may be used. The individual times are calculated, with their summation giving the total travel time. In large watersheds greater than 2000 acres, the stream flow velocities under the flood flow conditions should be used to estimate the Tc. *(Courtesy: American Iron and Steel Institute.)*

TABLE 7-11   Summary of Runoff Estimation Methods

METHOD	SIZE OF DRAINAGE AREA	REQUIRED INFORMATION	VARIABLES	OUTPUT	APPLICATIONS
RATIONAL METHOD	$\leq$ 20 acres (APWA Spec. Rep. No. 43)	Land Cover Tc IDF Curves	Runoff Coefficient (C)	Peak Flows	Minor System Design
UNIT HYDRO-GRAPH	$\leq$1000 sq mi (only daily rainfall and average daily discharge) Up to 5000 sq mi (extensive records required)	Rainfall and Streamflow Records		Hydrograph	Flood Flows Major System Storage Volumes
SCS UNIT HYDRO-GRAPH	Up to 20 sq mi; if large watershed break down to 20 sq mi sections	Soil Type Rainfall Data Tc	Runoff Curve No. (CN) Accounts for Hydrological Abstractions	Hydrograph	Flood Flows Minor & Major System Storage Volumes
SCS TABULAR METHOD	Up to 20 sq mi; if large watershed break down to 20 sq mi sections	Soil Type Cumulative Rainfall Tc	Runoff Curve No. (CN) Accounts for Hydrological Abstractions	Hydrograph	Flood Flows Major System Storage Volumes
SCS GRAPHICAL METHOD	$\leq$ 20 sq mi	Soil Type Cumulative Rainfall	Runoff Curve No. (CN) Accounts for Hydrological Abstractions	Peak Flow	Flood Peaks Minor & Major System
COMPUTER PROGRAMS	Dependent on capacity of program	Technical Information of Minor System Rainfall	See Users Manual	Hydrographs	Trouble Shooting Design of Minor and Major Systems Storage Volumes
RATIONAL MASS INFLOW	$\leq$ 20 Acres	Landcover IDF Curves Mass Outflow Curve	Runoff Coefficient (C)	Storage Volume	Detention Facility Design
HYDRO-GRAPH METHOD (Volume Determin-ation)	Dependent on source of hydrograph	Inflow Hydrograph Approximate Out-flow Hydrograph		Storage Volume Outflow Hydro-graph	Detention Fa-cility Design Reservoir Routing
SCS GRAPHICAL VOLUME METHOD	Dependent on source of peak flow	Peak Inflow Rate Desired Outflow Rate Type of control		Storage Volume	Detention Facility Design

SOURCE: American Iron and Steel Institute.

7-2 Chezy Equation

A basic hydraulic formula developed by Chezy for determining the flow of water particularly in open channels is written as follows:

$$Q = AV \qquad V = c\sqrt{RS} \qquad Q = Ac\sqrt{RS}$$

where Q = discharge, cubic feet per second
A = cross-sectional area of flow in square feet, at right angles to the direction of flow
V = mean velocity of flow, cubic feet per second
c = a coefficient of roughness whose value depends upon the character of surface over which water is flowing
R = hydraulic radius in feet = A/WP
WP = wetted perimeter or length, in feet, of wetted contact between a stream of water and its containing channel measured at right angles to the direction of flow
S = slope, or grade, in ft/ft

This fundamental formula is the basis of most capacity formulas.

This subsection courtesy of American Iron and Steel Institute.

7-3 Manning's Equation

Manning's formula, published in 1890, gives the value of c in the Chezy formula as

$$c = \frac{1.486}{n} R^{1/6}$$

the complete Manning formula being

$$V = \frac{1.486}{n} = R^{2/3} S^{1/2}$$

and combining with the Chezy equation

$$Q = A \, \frac{1.486}{n} \, R^{2/3} S^{1/2}$$

where A = cross-sectional area of flow in square feet
   S = slope, ft/ft
   R = hydraulic radius, feet
   n = coefficient of roughness

In many computations, it is convenient to group the properties peculiar to the cross section in one term called conveyance (K) or

$$K = \frac{1.486}{n} \, AR^{2/3}$$

Then $$Q = KS^{1/2}$$

Uniform flow of clean water in a straight, unobstructed channel would be a simple problem but is rarely attained. Manning's formula gives reliable results if the channel cross section, roughness, and slope are fairly constant over a sufficient distance to establish uniform flow.

This subsection courtesy of American Iron and Steel Institute.

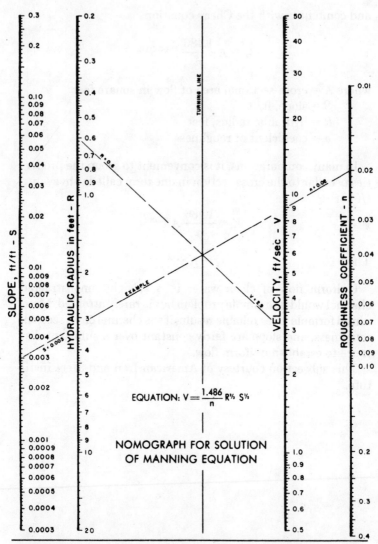

FIGURE 7-36 Nomograph for solution of Manning's equation. *(Courtesy: American Iron and Steel Institute.)*

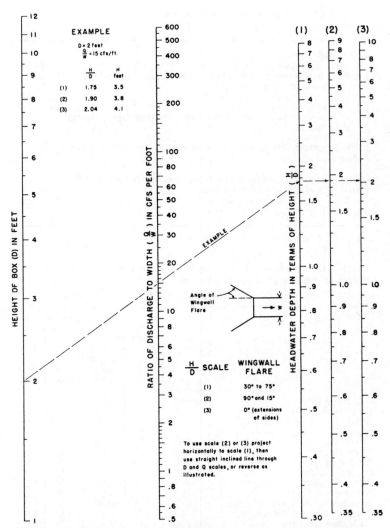

FIGURE 7-37 Nomograph for headwater depth of box culverts with entrance control. *(Source: U.S. Bureau of Public Roads.)*

TABLE 7-12   Manning's _n_ for Natural Stream Channels

Surface width at flood stage less than 100 feet.

1. Fairly regular section:
   a. Some grass and weeds, little or no brush................................ 0.030—0.035
   b. Dense growth of weeds, depth of flow materially greater than weed height.. 0.035—0.05
   c. Some weeds, light brush on banks.. 0.035—0.05
   d. Some weeds, heavy brush on banks....................................... 0.05 —0.07
   e. Some weeds, dense willows on banks..................................... 0.06 —0.08
   f. For trees within channel, with branches submerged at high stage, increase all
      above values by... 0.01 —0.02

2. Irregular sections, with pools, slight channel meander; increase values given
   above about... 0.01 —0.02

3. Mountain streams, no vegetation in channel, banks usually steep, trees and
   brush along banks submerged at high stage:
   a. Bottom of gravel, cobbles, and few boulders............................ 0.04 —0.05
   b. Bottom of cobbles, with large boulders................................. 0.05 —0.07

SOURCE: American Iron and Steel Institute.

TABLE 7-13   Comparison of Limiting Water Velocities and Tractive Force Values for the Design of Stable Channels

Straight channels after aging; canal depth, 3 feet.

Material	_n_	For Clear Water		Water Transporting Colloidal Silts	
		Velocity ft/sec	Tractive* Force lb/ft²	Velocity ft/sec	Trac-tive* Force lb/ft²
Fine sand colloidal	0.020	1.50	0.027	2.50	0.075
Sandy loam noncolloidal	0.020	1.75	0.037	2.50	0.075
Silt loam noncolloidal	0.020	2.00	0.048	3.00	0.11
Alluvial silts noncolloidal	0.020	2.00	0.048	3.50	0.15
Ordinary firm loam	0.020	2.50	0.075	3.50	0.15
Volcanic ash	0.020	2.50	0.075	3.50	0.15
Stiff clay very colloidal	0.025	3.75	0.26	5.00	0.46
Alluvial silts colloidal	0.025	3.75	0.26	5.00	0.46
Shales and hardpans	0.025	6.00	0.67	6.00	0.67
Fine gravel	0.020	2.50	0.075	5.00	0.32
Graded loam to cobbles when non-colloidal	0.030	3.75	0.38	5.00	0.66
Graded silts to cobbles when colloidal	0.030	4.00	0.43	5.50	0.80
Coarse gravel noncolloidal	0.025	4.00	0.30	6.00	0.67
Cobbles and shingles	0.035	5.00	0.91	5.50	1.10

*Tractive force or shear is the force which the water exerts on the periphery of a channel due to the motion of the water. The tractive values shown were computed from velocities given by S. Fortier and Fred C. Scobey and the values of _n_ shown.

The tractive force values are valid for the given materials regardless of depth. For depths greater than 3 ft. higher velocities can be allowed and still have the same tractive force.

From _U.S. Bureau of Reclamation_, Report No. Hyd-352, 1952, 60 pp.

SOURCE: American Iron and Steel Institute.

TABLE 7-14 Maximum Permissible Velocities in Vegetal-Lined Channels

Cover Average, Uniform Stand, Well Maintained	Slope Range	Permissible Velocity[a]	
		Erosion Resistant Soils	Easily Eroded Soils
	Percent	ft/sec	ft/sec
Bermudagrass	0–5 5–10 over 10	8 7 6	6 5 4
Buffalograss Kentucky bluegrass Smooth brome Blue grama	0–5 5–10 over 10	7 6 5	5 4 3
Grass mixture[b]	0–5 5–10	5 4	4 3
Lespedeza sericea Weeping lovegrass Yellow bluestem Kudzu Alfalfa Crabgrass	0–5	3.5	2.5
Common lespedeza[b] Sudangrass[b]	0–5[c]	3.5	2.5

[a]From *"Handbook of Channel Design for Soil and Water Conservation,"* Soil Conservation Service SCS-TP-61, Revised June 1954
[b]Annuals—used on mild slopes or as temporary protection until permanent covers are established
[c]Use on slopes steeper than 5 percent is not recommended.

SOURCE: American Iron and Steel Institute.

7-4 Channel Protection

If the mean velocity exceeds that permissible for the particular kind of soil, the channel should be protected from erosion. Grass linings are valuable where grass can be supported. Ditch bottoms may be sodded or seeded with the aid of temporary quick-growing grasses, mulches, jute bagging, or fiberglass. Grass may also be used in combination with other, more rigid types of linings, the grass being on the upper bank slopes.

Linings may consist of stone—dumped, hand placed, or grouted, preferably laid on a filter blanket of gravel or crushed stone.

Asphalt and concrete-lined channels are used on many steep erodible channels.

Corrugated steel flumes or chutes (and pipe spillways) are favored especially in wet, unstable, or frost-heaving soils. They should be anchored to prevent undue shifting. Most types of fabricated or poured channels should be protected against buoyancy and uplift, especially when empty. Cutoff walls, half diaphragms, or collars are used to prevent undermining.

Ditch checks (slope-control structures) are used in arid and semi-arid areas where grass won't grow. However, where grass will grow, the use of ditch checks in roadway or toe-of-slope channels is discouraged because they are a hazard to traffic and an impediment to mowing equipment.

High velocity at channel exits must be considered and some provision made to dissipate the excess energy.

This subsection courtesy of American Iron and Steel Institute.

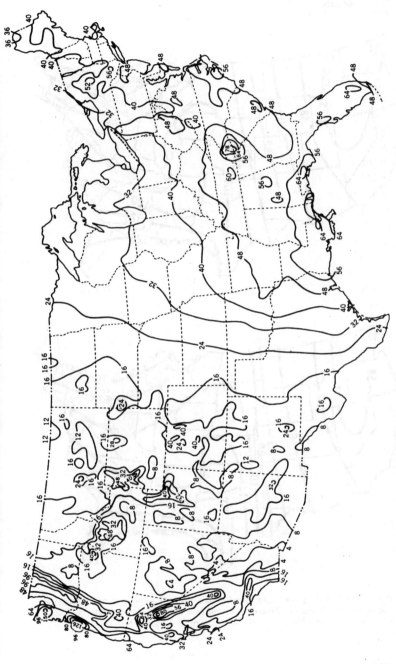

FIGURE 7-38 Average annual precipitation in United States. (*Source: National Weather Service.*)

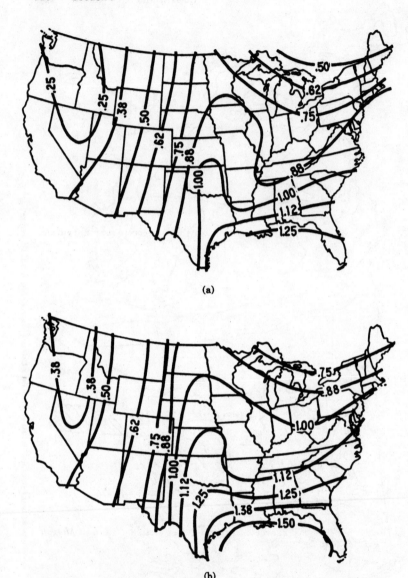

FIGURE 7-39 Fifteen-minute rainfall (in inches) expected once in (a) 2 years and (b) 5 years. *(Source: Rainfall Intensity-Frequency Data, Miscellaneous Publications No. 204, U.S. Department of Agriculture.)*

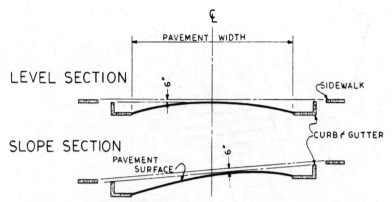

FIGURE 7-40 Typical cross section of street with concrete curb and gutter. *(Courtesy: Morgan & Parmley, Ltd.)*

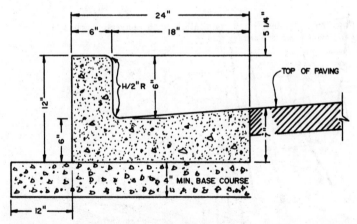

FIGURE 7-41 Typical concrete curb-and-gutter section. *(Courtesy: Morgan & Parmley, Ltd.)*

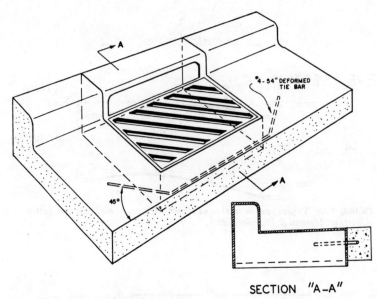

SECTION "A-A"

FIGURE 7-42 Typical inlet installation with concrete curb and gutter.
(Courtesy: Morgan & Parmley, Ltd.)

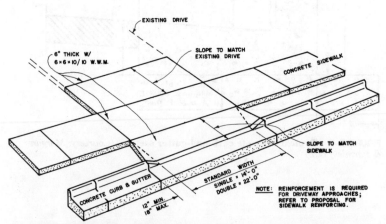

FIGURE 7-43 Typical driveway approach in concrete curb and gutter.
(Courtesy: Morgan & Parmley, Ltd.)

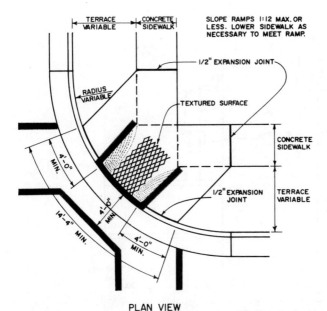

PLAN VIEW

FIGURE 7-44   Ramp detail at radius of concrete curb and gutter. *(Courtesy: Morgan & Parmley, Ltd.)*

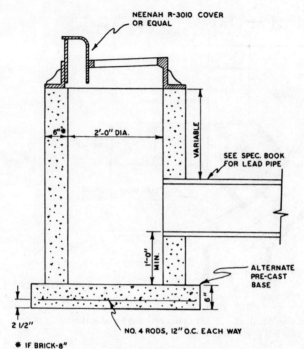

FIGURE 7-45 **Standard catch basin for storm sewer.** *(Courtesy: Morgan & Parmley, Ltd.)*

NOTES

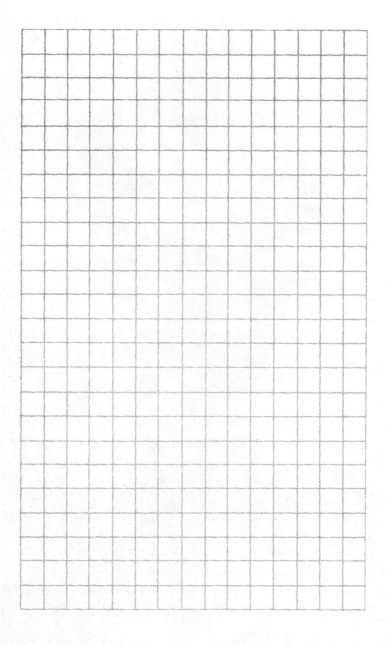

Storage and Fire Protection

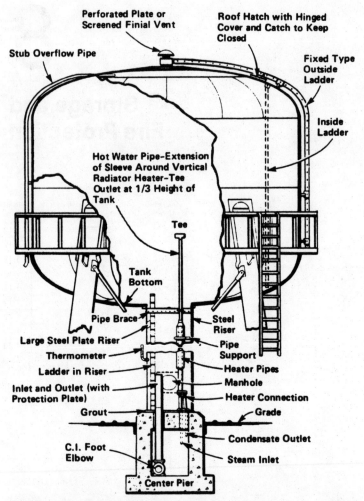

FIGURE 8-1 Typical gravity tank installation. *(Reprinted with permission from Automatic Sprinkler Systems Handbook, 2d ed., copyright 1985, National Fire Protection Assoc., Quincy, Mass.)*

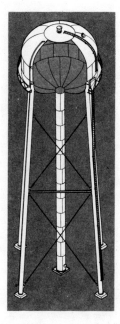

Capacity U.S. Gallons	Diameter Feet	Head Range Feet	Number of Columns	Radius Bevel
50,000	22.0	19.50	4	1:12
60,000	26.0	17.50	4	1:12
75,000	26.0	21.50	4	1:12
100,000	30.0	21.37	4	1:12
125,000	32.0	24.0	4	1:12
150,000	34.0	25.0	4	1:12
200,000	36.0	29.50	4	1:12
250,000	40.0	31.0	5	1:20
300,000	44.0	31.0	5	1:20

Capacity M³ Cubic Meters	Diameter Meters	Head Range Meters	Number of Columns	Radius Bevel
194	6.71	5.95	4	1:12
232	7.92	5.34	4	1:12
290	7.92	6.56	4	1:12
388	9.14	6.52	4	1:12
484	9.75	7.32	4	1:12
581	10.36	7.62	4	1:12
775	10.97	8.99	4	1:12
969	12.19	9.45	5	1:20
1,162	13.41	9.45	5	1:20

FIGURE 8-2 Double ellipsoidal elevated tank. *(Courtesy: Pitt-Moines, Inc.)*

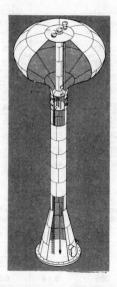

Pedestal Spheres

Capacity U.S. Gallons	Diameter Feet	Head Range Feet		Capacity M³ Cubic Meters	Diameter Meters	Head Range Meters
50,000	23.5	23.0		194	7.16	7.01
60,000	25.0	23.93		232	7.62	7.29
75,000	27.0	26.0		290	8.23	7.92
100,000	29.5	29.18		388	8.99	8.89
125,000	31.75	31.75		484	9.68	9.68
150,000	33.75	33.5		581	10.29	10.21
200,000	37.17	37.0		775	11.33	11.28

Pedestal Spheroids

Capacity U.S. Gallons	Diameter Feet	Head Range Feet		Capacity M³ Cubic Meters	Diameter Meters	Head Range Meters
200,000	40.0	30.0		775	12.19	9.14
250,000	45.0	30.0		969	13.71	9.14
300,000	47.25	30.0		1,162	14.40	9.14
400,000	53.0	35.0		1,550	15.85	10.67
400,000	53.0	30.0		1,550	16.15	9.14
500,000	55.5	37.5		1,938	16.92	11.43
500,000	58.75	30.0		1,938	17.91	9.14
750,000	65.0	40.0		2,907	19.81	12.19
1,000,000	74.0	40.0		3,875	22.56	12.19
1,000,000	77.0	35.0		3,875	23.47	10.67
1,500,000	85.25	46.0		5,813	25.98	14.02

FIGURE 8-3 Pedestal sphere elevated tank. *(Courtesy: Pitt-Moines, Inc.)*

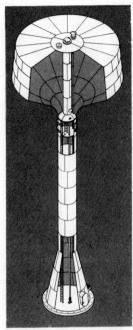

Capacity U.S. Gallons	Diameter Feet	Head Range Feet	Pillar Dia. Feet
150,000	38.0	28.5	8.0
200,000	43.0	30.0	8.0
250,000	48.0	30.0	10.0
300,000	51.0	32.0	10.0
400,000	55.0	35.0	12.0
500,000	60.0	37.5	12.0

Capacity M³ Cubic Meters	Diameter Meters	Head Range Meters	Pillar Dia. Meters
581	11.58	8.69	2.44
775	13.11	9.14	2.44
969	14.63	9.14	3.05
1,162	15.54	9.75	3.05
1,550	16.76	10.67	3.66
1,938	18.29	11.43	3.66

FIGURE 8-4 **Hydroped elevated tank.** *(Courtesy: Pitt-Moines, Inc.)*

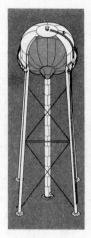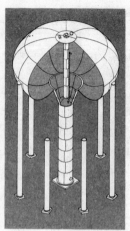

Torospherical (Vertical Shell)

Capacity U.S. Gallons	Diameter Feet	Head Range Feet	Number of Columns	Capacity M³ Cubic Meters	Diameter Meters	Head Range Meters	Number of Columns
200,000	36.0	29.67	4	775	10.97	9.04	4
250,000	40.0	28.42	5	969	12.19	8.66	5
300,000	45.0	28.08	5	1,162	13.72	8.56	5
400,000	48.0	32.36	6	1,550	14.63	9.86	6
500,000	50.0	37.33	6	1,938	15.24	11.38	6
500,000	56.0	29.85	6	1,938	17.07	9.10	6
750,000	62.5	35.0	8	2,907	19.05	10.67	8

Torospherical

Capacity U.S. Gallons	Diameter Feet	Head Range Feet	Number of Columns	Capacity M³ Cubic Meters	Diameter Meters	Head Range Meters	Number of Columns
500,000	62.25	25.0	7	1,938	18.97	7.62	7
750,000	70.17	30.0	8	2,907	21.39	9.14	8
750,000	75.17	25.0	8	2,907	22.91	7.62	8
1,000,000	75.0	35.0	9	3,875	22.86	10.67	9
1,000,000	80.0	30.0	9	3,875	24.38	9.14	9
1,000,000	86.0	25.0	10	3,875	26.21	7.62	10
1,500,000	91.0	35.0	12	5,813	27.74	10.67	12
1,500,000	97.0	30.0	12	5,813	29.57	9.14	12
2,000,000	105.0	34.75	14	7,750	32.00	10.59	14
2,500,000	108.0	40.0	16	9,689	32.92	12.19	16
3,000,000	126.5	35.0	16	11,626	38.56	10.67	16

FIGURE 8-5   Torospherical elevated tank. *(Courtesy: Pitt-Moines, Inc.)*

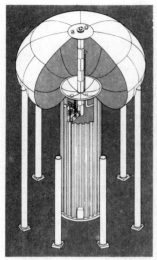

Capacity U.S. Gallons	Diameter Feet	Head Range Feet	Pillar Dia. Feet	Number of Columns
750,000	70.17	30.0	30.0	8
1,000,000	80.0	30.0	26.0	9
1,500,000	91.0	35.0	40.0	12
2,000,000	105.0	34.75	42.0	14
2,500,000	108.0	40.0	36.17	16
3,000,000	126.5	35.0	62.5	16

Capacity M³ Cubic Meters	Diameter Meters	Head Range Meters	Pillar Dia. Meters	Number of Columns
2,907	21.39	9.14	9.14	8
3,875	24.38	9.14	7.92	9
5,813	27.74	10.67	12.19	12
7,750	32.00	10.59	12.80	14
9,689	32.92	12.19	11.02	16
11,626	38.56	10.67	19.05	16

FIGURE 8-6 Toropillar elevated tank. *(Courtesy: Pitt-Moines, Inc.)*

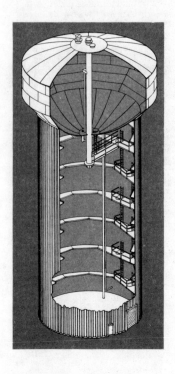

Capacity U.S. Gallons	Diameter Feet	Head Range Feet	Pillar Dia. Feet
200,000	42.0	25.0	24.0
250,000	42.0	30.0	24.0
300,000	44.0	33.0	24.0
400,000	44.0	39.0	24.0
500,000	49.5	38.0	30.0
750,000	64.0	40.0	42.0
1,000,000	˙74.0	40.0	52.0
ˌ1,500,000	86.0	42.5	60.0
2,000,000	100.0	40.0	78.0
2,500,000	108.0	44.0	78.0
3,000,000	120.0	42.0	90.0

Capacity M³ Cubic Meters	Diameter Meters	Head Range Meters	Pillar Dia. Meters
775	12.80	7.62	7.32
969	12.80	9.14	7.32
1,162	13.41	10.06	7.32
1,550	13.41	11.89	7.32
1,938	15.09	11.58	9.14
2,907	19.51	12.19	13.41
3,875	22.56	12.19	15.85
5,813	26.21	12.95	18.29
7,750	30.48	12.19	23.77
9,689	32.92	13.41	23.77
11,626	36.58	12.80	27.43

FIGURE 8-7 Hydropillar elevated tank. *(Courtesy: Pitt-Moines, Inc.)*

Capacity U.S. Gallons	Diameter Feet	Head Range Feet	Pillar Dia. Feet
100,000	36	23.0	14
150,000	37	29.0	14
200,000	39	35.0	14
250,000	40	40.17	14
300,000	43	39.5	18
400,000	49	39.33	24

Capacity M³ Cubic Meters	Diameter Meters	Head Range Meters	Pillar Dia. Meters
388	10.97	7.01	4.27
581	11.28	8.84	4.27
775	11.89	10.67	4.27
969	12.19	12.24	4.27
1,162	13.11	12.04	5.49
1,550	14.94	11.98	7.32

FIGURE 8-8 Hydropillar elevated tank (wineglass style). *(Courtesy: Pitt-Moines, Inc.)*

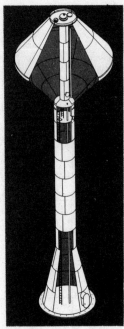

Capacity U.S. Gallons	Diameter Feet	Head Range Feet	Pedestal Dia. Feet
50,000	29.08	20.0	6.5
75,000	33.25	24.0	6.5
100,000	36.67	26.67	6.5

Note: Smaller capacities also available upon request.

Capacity M³ Cubic Meters	Diameter Meters	Head Range Meters	Pedestal Dia. Meters
194	8.86	6.1	1.98
290	10.13	7.32	1.98
388	11.18	8.13	1.98

Note: Smaller capacities also available upon request.

FIGURE 8-9 Double-cone elevated tank. *(Courtesy: Pitt-Moines, Inc.)*

TABLE 8-1 Guide to Water Supply Requirements for Pipe Schedule Sprinkler Systems

Occupancy Classification	Residual Pressure Required (see Note 1)	Acceptable Flow at Base of Riser (see Note 2)	Duration in Minutes (see Note 4)
Light Hazard	15 psi	500-750 gpm (see Note 3)	30-60
Ordinary Hazard (Group 1)	15 psi or higher	700-1000 gpm	60-90
Ordinary Hazard (Group 2)	15 psi or higher	850-1500 gpm	60-90
Ordinary Hazard (Group 3)	Pressure and flow requirements for sprinklers and hose streams to be determined by authority having jurisdiction.		60-120
High-Piled Storage	Pressure and flow requirements for sprinklers and hose streams to be determined by authority having jurisdiction.		
High-Rise Buildings	Pressure and flow requirements for sprinklers and hose streams to be determined by authority having jurisdiction.		
Extra Hazard	Pressure and flow requirements for sprinklers and hose streams to be determined by authority having jurisdiction.		

For SI Units: 1 psi = 0.0689 bar; 1 gpm = 3.785 L/min.

Notes:

1. The pressure required at the base of the sprinkler riser(s) is defined as the residual pressure required at the elevation of the highest sprinkler plus the pressure required to reach this elevation.

2. The lower figure is the minimum flow including hose streams ordinarily acceptable for pipe schedule sprinkler systems. The higher flow should normally suffice for all cases under each group.

3. The requirement may be reduced to 250 gpm if building area is limited by size or compartmentation or if building (including roof) is noncombustible construction.

4. The lower duration figure is ordinarily acceptable where remote station water flow alarm service or equivalent is provided. The higher duration figure should normally suffice for all cases under each group.

SOURCE: Reprinted with permission from *Automatic Sprinkler Systems Handbook*, 2d ed. Copyright 1985, National Fire Protection Assoc., Quincy, Mass.

TABLE 8-2   Table and Design Curves for Minimum Water Supply and Density

Hazard Classification	Sprinklers Only — gpm	Inside Hose — gpm	Total Combined Inside and Outside Hose — gpm	Duration in Minutes
Light	See 2-2.1.2.1	0, 50 or 100	100	30
Ord. — Gp. 1	See 2-2.1.2.1	0, 50 or 100	250	60-90
Ord. — Gp. 2	See 2-2.1.2.1	0, 50 or 100	250	60-90
Ord. — Gp. 3	See 2-2.1.2.1	0, 50 or 100	500	60-120
Ex. Haz. — Gp. 1	See 2-2.1.2.1	0, 50 or 100	500	90-120
Ex. Haz. — Gp. 2	See 2-2.1.2.1	0, 50 or 100	1000	120

For SI Units: 1 gpm=3.785 L/min.

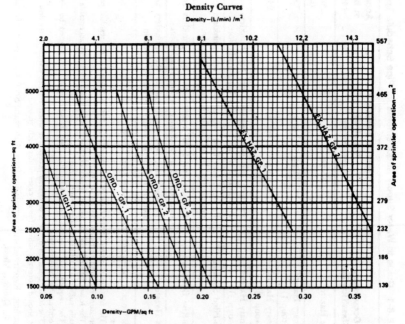

Density Curves

Density—(L/min) /m²

For SI Units: 1 sq ft = 0.0920 m ; 1 gpm/sq ft = 40.746 (L/min)/m

SOURCE: Reprinted with permission from *Automatic Sprinkler Systems Handbook*, 2d ed. Copyright 1985, National Fire Protection Assoc., Quincy, Mass.

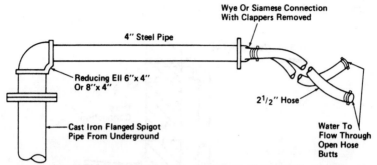

Employing Horizontal Run Of 4-Inch Pipe And Reducing Fitting Near Base Of Riser.

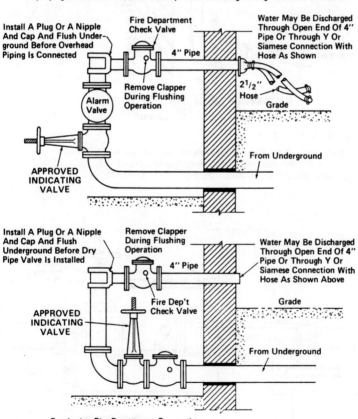

Employing Fire Department Connections.

For SI Units: 1 in. = 25.4 mm.

FIGURE 8-10 Methods of flushing water supply connections. *(Reprinted with permission from Automatic Sprinkler Systems Handbook, 2d ed., copyright 1985, National Fire Protection Assoc., Quincy, Mass.)*

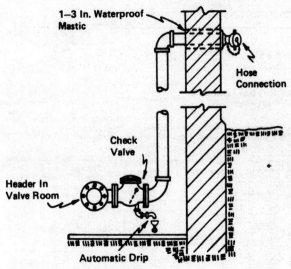

1–3 In. Waterproof
Mastic

Hose
Connection

Check
Valve

Header In
Valve Room

Automatic Drip

For SI Units: 1 in. = 25.4 mm.

FIGURE 8-11   Fire department connection. *(Reprinted with permission from Automatic Sprinkler Systems Handbook, 2d ed., copyright 1985, National Fire Protection Assoc., Quincy, Mass.)*

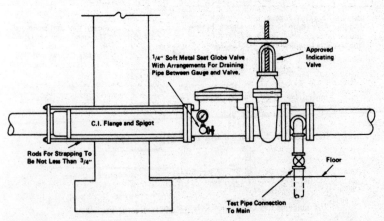

1/4" Soft Metal Seat Globe Valve
With Arrangements For Draining
Pipe Between Gauge and Valve.

Approved
Indicating
Valve

C.I. Flange and Spigot

Rods For Strapping To
Be Not Less Than 3/4"

Floor

Test Pipe Connection
To Main

For SI Units: 1 in. = 25.4 mm.

FIGURE 8-12   Water supply connection with test pipe. *(Reprinted with permission from Automatic Sprinkler Systems Handbook, 2d ed., copyright 1985, National Fire Protection Assoc., Quincy, Mass.)*

TABLE 8-3 Pressure and Number of Design Sprinklers for Various Hazards

See Note 6.

Minimum Operating Pressure (Note 1), psi (bar)	25 (1.7)	50 (3.4)	75 (5.2)
Hazard (Note 2)	Number Design Sprinklers		
Palletized Storage			
Class I, II and commodities up to 25 ft (7.6 m) with maximum 10 ft (3.0 m) clearance to ceiling	15	Note 3	Note 3
Class IV commodities up to 20 ft (6.1 m) with maximum 10 ft (3.0 m) clearance to ceiling	20	15	Note 3
Unexpanded plastics up to 20 ft (6.1 m) with maximum 10 ft (3.0 m) clearance to ceiling	25	15	Note 3
Idle wood pallets up to 20 ft (6.1 m) with maximum 10 ft (3.0 m) clearance to ceiling	15	Note 3	Note 3
Solid-Piled Storage			
Class I, II and III commodities up to 20 ft (6.1 m) with maximum 10 ft (3.0 m) clearance to ceiling	15	Note 3	Note 3
Class IV commodities and unexpanded plastics up to 20 ft (6.1 m) with maximum 10 ft (3.0 m) clearance to ceiling	Does Not Apply	15	Note 3
Double-Row Rack Storage with Minimum 5½ ft (1.7 m) Aisle Width (Note 4)			
Class I and II commodities up to 25 ft (7.6 m) with maximum 5 ft (1.5 m) clearance to ceiling	20	Note 3	Note 3
Class I, II, and III commodities up to 20 ft (6.1 m) with maxiumum 10 ft (3.0 m) clearance to ceiling	15	Note 3	Note 3
Class IV commodities up to 20 ft (6.1 m) with maximum 10 ft (3.0 m) clearance to ceiling	Does Not Apply	20	15
Unexpanded plastics up to 20 ft (6.1 m) with maximum 10 ft (3.0 m) clearance to ceiling	Does Not Apply	30	20
Unexpanded plastics up to 20 ft (6.1 m) with maximum 10 ft (3.0 m) clearance to ceiling (Note 7)	Does Not Apply	20	Note 3
Class IV commodities and unexpanded plastics up to 20 ft (6.1 m) with maximum 5 ft (1.5 m) clearance to ceiling	Does Not Apply	15	Note 3

TABLE 8-3   Pressure and Number of Design Sprinklers for Various Hazards *(Continued)*

On-End Storage of Roll Paper (Note 5)

Heavyweight paper, in closed array, banded in open array, or banded or unbanded in a standard array, up to 26 ft (7.9 m) with a maximum 34 ft (10.4 m) clearance to ceiling	Does Not Apply	15	Note 3
Any grade of paper, EXCEPT LIGHTWEIGHT paper, with stacks in closed array, or banded or unbanded standard array up to 20 ft (6.1 m) with maximum 10 ft (3.0 m) clearance to ceiling	Does Not Apply	15	Note 3
Medium weight paper completely wrapped (sides and ends) in one or more layers of heavyweight paper, or lightweight paper in two or more layers of heavyweight paper with stacks in closed array, banded in open array, or banded or unbanded in a standard array, up to 26 ft (7.9 m) with maximum 34 ft (10.4 m) clearance to ceiling	Does Not Apply	15	Note 3

Record Storage

Paper records and/or computer tapes in multitier steel shelving up to 5 ft (1.5 m) in width and with aisles 30 in. (76 cm) or wider, without catwalks in the aisles, up to 15 ft (4.6 m) with maximum 5 ft (1.5 m) clearance to ceiling	15	Note 3	Note 3
Same as above, but with catwalks of expanded metal or metal grid with minimum 50 percent open area, in the aisles	Does Not Apply	15	Note 3

Notes:

1. Open Wood Joist Construction. Testing with open wood joist construction showed that each joist channel should be fully firestopped to its full depth at intervals not exceeding 20 ft (6.1 m). In unfirestopped open wood joist construction, or if firestops are installed at intervals exceeding 20 ft (6.1 m), the minimum operating pressures should be increased by 40 percent.

2. Building steel required no special protection for the occupancies listed.

3. The higher pressure will successfully control the fire, but the required number of design sprinklers should not be reduced from that required for the lower pressure.

4. In addition to the transverse flue spaces required by NFPA 231C, minimum 6 in. (152 mm) longitudinal flue spaces were maintained.

5. See NFPA 231F for definitions.

6. Unless otherwise specified the sprinklers used in the tests were high temperature rating.

7. Based on tests using sprinkler or ordinary temperature rating.

SOURCE: Reprinted with permission from *Automatic Sprinkler Systems Handbook*, 2d ed. Copyright 1985, National Fire Protection Assoc., Quincy, Mass.

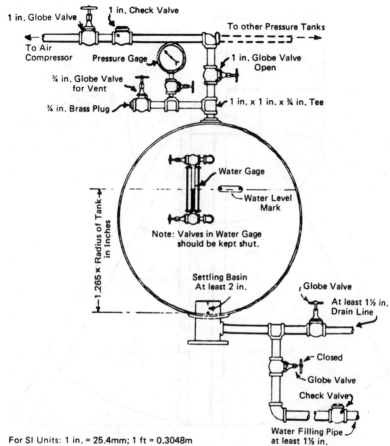

For SI Units: 1 in. = 25.4mm; 1 ft = 0.3048m

FIGURE 8-13 Typical connection to pressure tanks. *(Reprinted with permission from Automatic Sprinkler Systems Handbook, 2d ed., copyright 1985, National Fire Protection Assoc., Quincy, Mass.)*

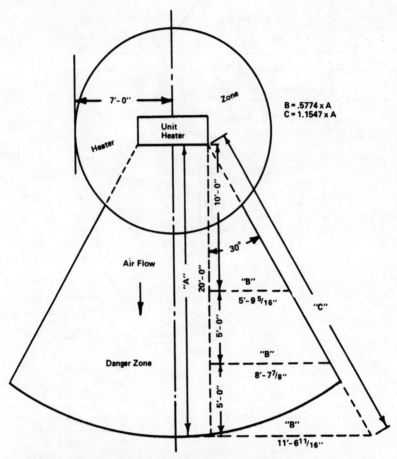

For SI Units: 1 in. = 25.4 mm; 1 ft = 0.3048 m.

FIGURE 8-14 Heater and danger zones at unit heaters. *(Reprinted with permission from Automatic Sprinkler Systems Handbook, 2d ed., copyright 1985, National Fire Protection Assoc., Quincy, Mass.)*

Replacement parts and dimensions for the Mueller Standard and Post-type fire hydrants

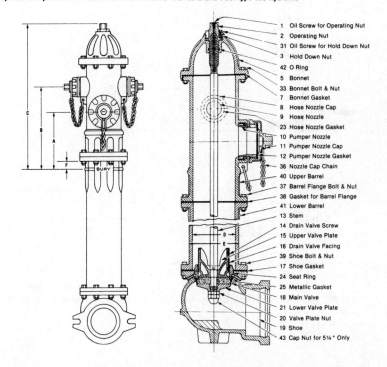

		1	Oil Screw for Operating Nut
		2	Operating Nut
		31	Oil Screw for Hold Down Nut
		3	Hold Down Nut
		42	O Ring
		5	Bonnet
		33	Bonnet Bolt & Nut
		7	Bonnet Gasket
		8	Hose Nozzle Cap
		9	Hose Nozzle
		23	Hose Nozzle Gasket
		10	Pumper Nozzle
		11	Pumper Nozzle Cap
		12	Pumper Nozzle Gasket
		36	Nozzle Cap Chain
		40	Upper Barrel
		37	Barrel Flange Bolt & Nut
		38	Gasket for Barrel Flange
		41	Lower Barrel
		13	Stem
		14	Drain Valve Screw
		15	Upper Valve Plate
		16	Drain Valve Facing
		39	Shoe Bolt & Nut
		17	Shoe Gasket
		24	Seat Ring
		25	Metallic Gasket
		18	Main Valve
		21	Lower Valve Plate
		20	Valve Plate Nut
		19	Shoe
		43	Cap Nut for 5¼″ Only

Dimensions

Hydrant size	A*		B		C		D		E		F	
	inch	mm	inch	mm	inch	mm	inch	mm	inch	mm	inch	mm
2 1/8″			14 1/2″	368	23 3/4″	603	3 1/4″	83	—	—	2 5/8″	67
4 1/4″	12 1/4″	311	17 3/4″	451	30 3/4″	781	6″	152	7″	178	1 3/8″	35
5 1/4″	12 1/4″	311	17 3/4″	451	31 3/4″	806	7″	178	8 1/8″	206	1 3/8″	35

FIGURE 8-15 Details of fire hydrant. *(Courtesy: Mueller Company, Decatur, Ill.)*

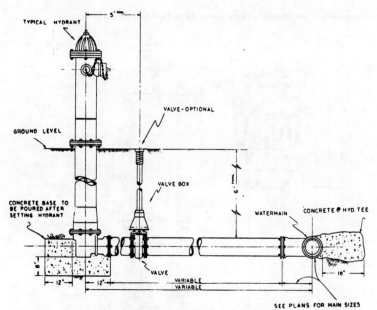

FIGURE 8-16 Typical hydrant installation. *(Courtesy: Morgan & Parmley, Ltd.)*

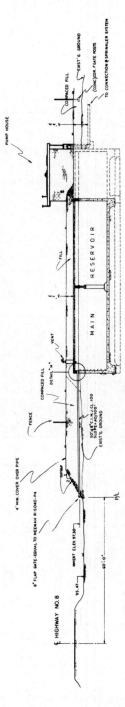

FIGURE 8-17 Subsurface fire protection facility. *(Courtesy: Morgan & Parmley, Ltd.)*

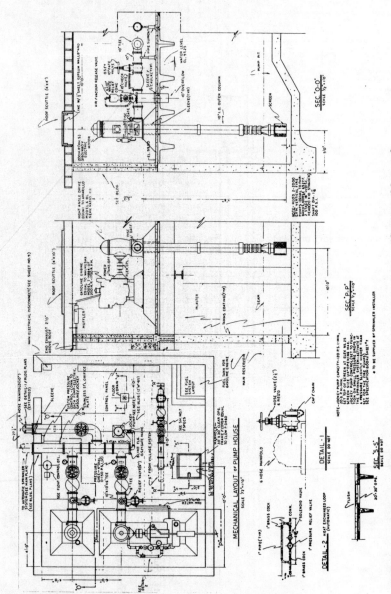

FIGURE 8-18 Pumphouse details of subsurface fire protection facility. *(Courtesy: Morgan & Parmley, Ltd.)*

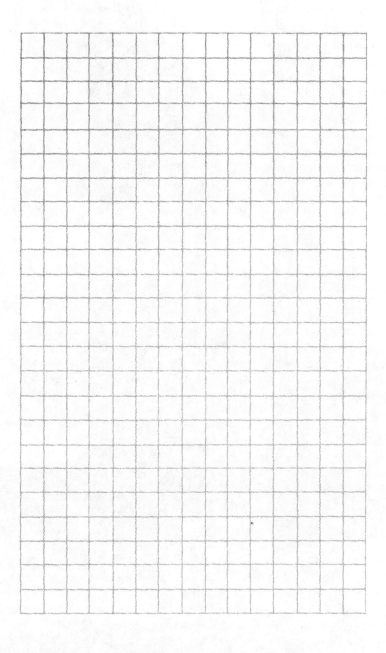

Estimating
Flows
in the Field

9-1   Field Calculations

Each user of this manual certainly will have different needs and areas of varying technical reference. However, this manual's primary function is to provide technical assistance in the "field." Therefore, this section contains some basic data directed toward field calculations, most commonly encountered by the editor, but not necessarily all-inclusive to every user. It is strongly recommended that the user highlight specific pages throughout this manual that are most useful and record supplemental data on the appropriate blank Notes pages which follow all sections.

It should be further noted that the contents of this and other sections are subordinate to applicable regulatory codes and site-specific guidelines.

The user should also refer to available instruction manuals prepared by manufacturers of measuring and recording equipment that one may encounter on a specific field reconnaissance mission.

9-2   Density of Water

The density of water varies relative to temperature. Its maximum density occurs at 39.2°F (4°C).

Fresh water weighs 62.425 pounds per cubic foot or 1000 kilograms per cubic meter.

9-3   Water Pressure

The total pressure exerted by the weight of a water column is determined by the column's height. The configuration of the column does not affect the pressure.

Net or normal pressure is the force applied to the pipe wall by a liquid in said pipe. Without flow in the pipe, this condition is known as "static" pressure.

The force or pressure generated by a column of water 1 foot high $= 62.4/144 = w = 0.433$ psi. The head (h), in vertical feet, corresponding to a pressure p (psi) will be $h = p/w = p/0.433$.

9-4   Archimedes' Principle

Archimedes' principle states that an object placed in water seems to lose an amount of weight equal to the weight of the water it displaces. Archimedes, a Greek mathematician, developed this principle in the second century B.C.

9-5   Pascal's Law

A fluid in a container transmits pressure equally in all directions. Blaise Pascal, a French scientist and mathematician, developed this law during the seventeenth century A.D.

9-6   Continuity in Fluid Flow

The principle of continuity in fluid flow states that the velocity of water flowing through a pipe (conduit) increases as the cross-sectional area of the pipe decreases, and decreases as the pipe's cross-sectional area increases.

9-7   Bernoulli's Law

The pressure of a liquid (or water) increases as its velocity decreases and decreases as the velocity of the liquid (or water) increases. Daniel Bernoulli was a Swiss mathematician who lived during the eighteenth century A.D.

9-8   Velocity Head

Velocity head or velocity pressure generated by a volume or mass of water by pressure action equals free fall of said water mass, beginning from rest, through a given distance equivalent to the head pressure in feet, or $v = \sqrt{2gh}$, where v is expressed in feet per second.

9-9   Orifice Flow

The flow rate Q or discharge through an orifice may be expressed in terms of velocity v and cross-sectional area a of the stream. Assuming orifice area in square feet (a), discharge in cubic feet per second (Q), and head (h) in feet, the following formula may be used:

$$Q = ca\sqrt{2gh}$$

where c = roughness coefficient
g = acceleration of gravity = 32.16

9-10 Friction Loss Flow

The accepted expression relating velocity to friction loss in pipe is the Chezy formula:

$$v = c\sqrt{rs}$$

where v = velocity
c = roughness of pipe
r = hydraulic radius (area/circumference)
s = hydraulic slope [h/L or (friction head/length of pipe)]

9-11 Water Hammer

The effect of pressure increase that may accompany a sudden change in water velocity in a pipe network is known as *water hammer*. If the deceleration of water velocity is quick or totally stopped, the kinetic force of the water column is momentarily absorbed by the elasticity of the pipe wall and compressibility of the water itself. This surge force may sometimes be strong enough to rupture the piping system. For this reason, valves, hydrants, and flow-control units should always be closed or opened "slowly" to avoid development of the water hammer effect. The initial shock wave produces the maximum pressure, and each succeeding shock lessens in intensity. The time lapse between individual shocks is proportional to the pipe length affected and wave velocity. Severe water hammer is a very serious condition and may result in the failure of the piping system.

9-12 Hydraulic Press Principle

The hydraulic press principle increases and transfers the force applied to a small piston through a liquid to a larger piston. This force increases in direct proportion to the diameter, or cross-sectional area, of the larger piston.

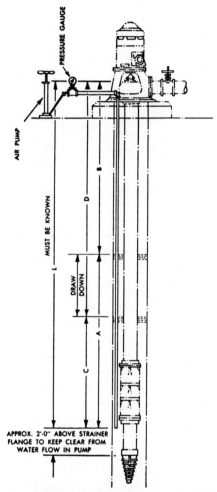

FIGURE 9-1 Method of determining the depth to water level in a deep well.
In testing a vertical submerged pump such as a deep well turbine, it is necessary
to determine the water level in the well when pumping. The most satisfactory
method of determining the water level involves the use of a ¼-inch air line of
known vertical length, a pressure gauge, and an ordinary bicycle or automobile
pump installed as shown in this figure. If possible, the air line pipe should reach
at least 20 feet beyond the lowest anticipated water level in the well in order to
assure more reliable gauge readings and preferably should not be attached to the
column or bowls, as this would hinder the removal of the pipe should any leaks
develop. As noted in this figure, an air pressure gauge is used to indicate the
pressure in the air line.

(*Continued on next page*)

FIGURE 9-1 (*Continued*)

The ¼-inch air line pipe is lowered into the well, a tee is placed in the line above the ground, and a pressure gauge is screwed into one connection and the other is fitted with an ordinary bicycle valve to which a bicycle pump is attached. All joints must be made carefully and must be airtight to obtain correct information. When air is forced into the line by means of the tire pump the gauge pressure increases until all the water has been expelled. When this point is reached the gauge reading becomes constant. The maximum maintained air pressure recorded by the gauge is equivalent to that necessary to support a column of water of the same height as that forced out of the air line. The length of this water column is equal to the amount of air line submerged.

Deducting this pressure converted to feet (pounds pressure × 2.31 equals feet) from the known length of the ¼-inch air line pipe will give the amount of submergence. The following examples will serve to clarify the above explanation.

Assume a length L of 150 feet.

Pressure gauge reading before starting pump = P_1 = 25 pounds per square inch. Then $A = 25 \times 2.31 = 57.7$ feet; therefore the water level in the well before starting the pump would be $B = L - A = 150 - 57.7 = 92.3$ feet.

Pressure gauge reading when pumping = P_2 = 18 pounds per square inch. Then $C = 18 \times 2.31 = 41.6$ feet; therefore the water level in the well when pumping would be $D = L - C = 150 - 41.6$ feet = 108.4 feet.

The drawdown is determined by the following equation: $D - B = 108.4 - 92.3 = 16.1$ feet.

(Courtesy: Fairbanks Morse Pump Corporation.)

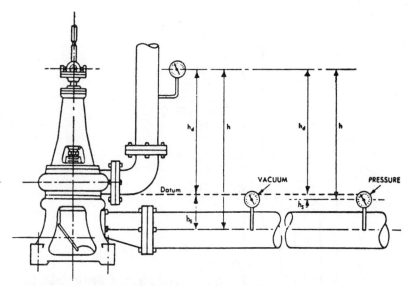

FIGURE 9-2 Determination of total head from gauge readings. In pump tests the total head can be determined by gauges as illustrated in this figure, where the total head would be determined as follows:

H = discharge gauge reading (corrected, feet liquid) + vacuum gauge reading (corrected, feet liquid) + distance between point of attachment of vacuum gauge to the center line of discharge gauge (h, feet) + ($V_d^2/2g - V_s^2/2g$), or

H = discharge gauge reading (corrected, feet liquid) − pressure gauge reading in suction line (corrected, feet liquid) + distance between center of discharge and center of suction gauges (h, feet) + ($V_d^2/2g - V_s^2/2g$).

The method of head determination above applies specifically to pumping units installed so that both suction and discharge flanges of the pump and adjacent piping are located so as to be accessible for installation of gauges for testing the pump.

(Courtesy: Fairbanks Morse Pump Corporation.)

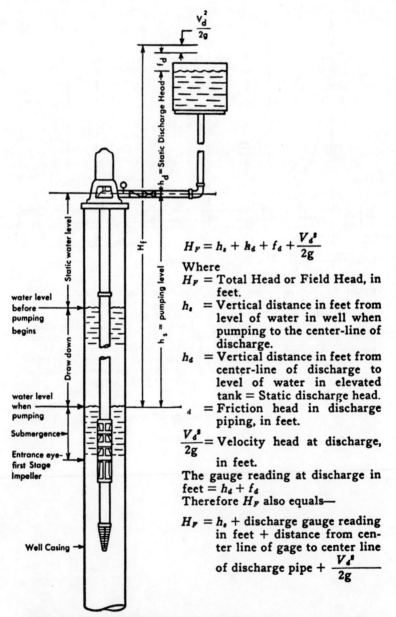

$$H_F = h_s + h_d + f_d + \frac{V_d^2}{2g}$$

Where

H_F = Total Head or Field Head, in feet.

h_s = Vertical distance in feet from level of water in well when pumping to the center-line of discharge.

h_d = Vertical distance in feet from center-line of discharge to level of water in elevated tank = Static discharge head.

f_d = Friction head in discharge piping, in feet.

$\dfrac{V_d^2}{2g}$ = Velocity head at discharge, in feet.

The gauge reading at discharge in feet = $h_d + f_d$
Therefore H_F also equals—

$H_F = h_s + $ discharge gauge reading in feet + distance from center line of gage to center line of discharge pipe + $\dfrac{V_d^2}{2g}$

FIGURE 9-3 Determination of total head of deep-well turbine or propeller pump. *(Courtesy: Fairbanks Morse Pump Corporation.)*

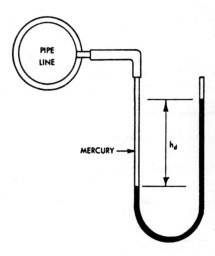

FIGURE 9-4 Manometer pressure calculation methods. If the space above the mercury in both legs of the manometer is filled with air, the pressure in the pipe line can be calculated as follows: H, feet water $= h_d$, inches mercury $\times$ 13.6/12 $= h_d \times 1.133$, where 13.6 = specific gravity of mercury. However, if the left-hand leg above the mercury is filled with water, the weight of the water, h_d, causes extra deflection of the mercury. In this case, therefore, it is necessary to subtract the specific gravity of water from the specific gravity of mercury in arriving at the head in the pipe; thus H, feet water $= h_d$, inches mercury $\times$ (13.6 − 1)/12 $= h_d \times$ 1.05. *(Courtesy: Fairbanks Morse Pump Corporation.)*

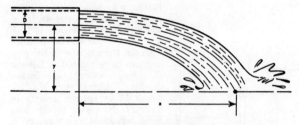

Approximating flow from horizontal pipe.

$$\text{Capacity, Gpm} = \frac{2.45\,D^2\,x}{\sqrt{\dfrac{2\,y}{32.16}}}$$

Where D = Pipe diameter, in.

x = Horizontal distance, ft.

y = Vertical distance, ft.

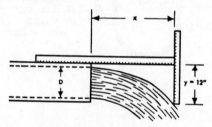

Approximating flow from horizontal pipe.

Capacity, Gpm = $0.818\,D^2X$

Std. Wt. Steel Pipe, Inside Dia., In.		Distance x, in., when y = 12"										
Nominal	Actual	12	14	16	18	20	22	24	26	28	30	32
2	2.067	42	49	56	63	70	77	84	91	98	105	112
2½	2.469	60	70	80	90	100	110	120	130	140	150	160
3	3.068	93	108	123	139	154	169	185	200	216	231	246
4	4.026	159	186	212	239	266	292	318	345	372	398	425
5	5.047	250	292	334	376	417	459	501	543	585	627	668
6	6.065	362	422	482	542	602	662	722	782	842	902	962
8	7.981	627	732	837	942	1047	1150	1255	1360	1465	1570	1675
10	10.020	980	1145	1310	1475	1635	1800	1965	2130	2290	2455	2620
12	12.000	1415	1650	1890	2125	2360	2595	2830	3065	3300	3540	3775

FIGURE 9-5 Approximating flow from horizontal pipes.
Often an approximation of water flow is required when it is not practical to use weirs, orifices, nozzles, or other means of determination. This can be done by taking the coordinates of a point in the stream flow as indicated. The accuracy of this method will vary from 90 to 100 percent. The pipe must be flowing full.

This can be further simplified by measuring to the top of the flowing stream and always measuring so that y will equal 12 inches and measuring the horizontal distance X in inches as illustrated.

(Courtesy: Fairbanks Morse Pump Corporation.)

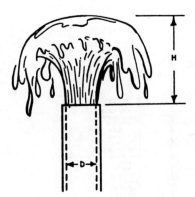

Capacity, Gpm. = 5.68 KD² H¹ᐟ²

D = I.D. of Pipe, In.

H = Vertical Height of water jets, in.

K = a constant, varying from .87 to .97 for pipes 2 to 6 in. dia. and H = 6 to 24 in.

FLOW FROM VERTICAL PIPES, GPM.

Nominal	Vertical Height, H, of Water Jet, in.										
I.D. Pipe, in.	3	3.5	4	4.5	5	5.5	6	7	8	10	12
2	38	41	44	47	50	53	56	61	65	74	82
3	81	89	96	103	109	114	120	132	141	160	177
4	137	151	163	174	185	195	205	222	240	269	299
6	318	349	378	405	430	455	480	520	560	635	700
8	567	623	684	730	776	821	868	945	1020	1150	1270
10	950	1055	1115	1200	1280	1350	1415	1530	1640	1840	2010

FIGURE 9-6 Approximating flow from vertical pipes. *(Courtesy: Fairbanks Morse Pump Corporation.)*

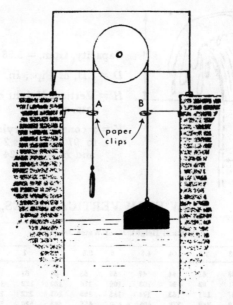

With a pair of paper clips and two brackets you can improvise an inexpensive high-low level indicator. The brackets can be either wood or metal. Affix the brackets to the stilling well wall as shown in the drawing. They must protrude far enough out to touch the float tape at points A and B and prevent the paper clips from rising above those points.

Place the paper clips on the float tape below the brackets. As the water level drops the clip at point A will slide towards the counter-weight. As water level rises the clip at point B will move nearer to the float.

To read the indicator for high level measure the tape from clip B to the float. A graduated tape will make this reading easy. This measurement is the highest water level below bracket B.

To measure low level determine the tape distance of clip A from the float. Subtract the length over the pulley between brackets A and B. This will give the low water level below bracket B.

FIGURE 9-7   High-low level indicator. *(Courtesy: Leupold & Stevens, Inc.)*

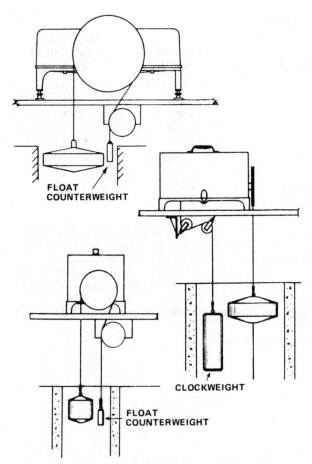

FIGURE 9-8 Illustrated uses of float assemblies. *(Courtesy: Leupold & Stevens, Inc.)*

TABLE 9-1 Float Well Sizing

Diameter of Float Well	Diameter of Inlet Hole	Diam. of Inlet Pipe 20 to 30 Ft. Long
12 inches	1/2 inch	1/2 inch
16 inches	1/2	3/4
20 inches	5/8	3/4
24 inches	3/4	1
30 inches	1	1-1/2
36 inches	1-1/4	2
3x3 feet square	1-1/4	2
3x4 feet rectangular	1-1/2	3
4x5 feet rectangular	2	4

Materials. Float wells may be made of wood or reinforced concrete, rectangular or square; of round metal pipe, cement or vitrified sewer pipe. Corrugated galvanized iron culvert pipe with seams soldered make excellent wells, since they will stand external pressure of earth or ice. The corrugations also serve for holding iron bands anchored into a cliff, pier or wall.

The well must have a bottom and be practically water tight except for the water inlet.

SOURCE: Leupold & Stevens, Inc.

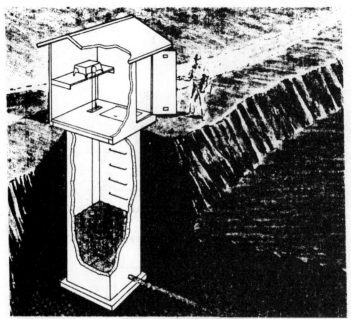

FIGURE 9-9 Typical river gauging station. *(Courtesy: Leupold & Stevens, Inc.)*

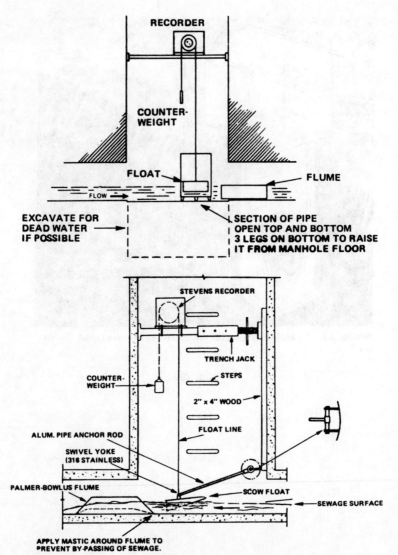

FIGURE 9-10 Details of recording stations using floats. *(Courtesy: Leupold & Stevens, Inc.)*

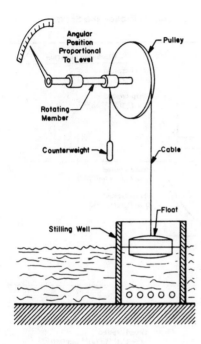

FIGURE 9-11 Details of float-operated flow meter. *(Courtesy: Isco, Inc.)*

TABLE 9-2   Summary of Discharge Equations for Broad- and Sharp-Crested Weirs

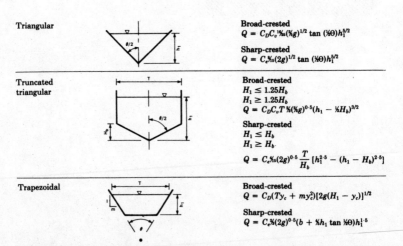

Type	Schematic definition of control section	Discharge equation
Rectangular		Broad-crested $Q = C_D C_v \frac{2}{3}(\frac{2}{3}g)^{1/2} T h_1^{3/2}$ Sharp-crested $Q = C_e \frac{2}{3}(2g)^{1/2} b h_1^{3/2}$
Parabolic		Broad-crested $Q = C_D C_v (\frac{2}{3}fg)^{1/2} h_1^2$ Sharp-crested $Q = C_e \frac{1}{2}\pi (fg)^{1/2} h_1^2$
Triangular		Broad-crested $Q = C_D C_v \frac{16}{25}(\frac{2}{5}g)^{1/2} \tan (\frac{1}{2}\Theta) h_1^{5/2}$ Sharp-crested $Q = C_e \frac{8}{15}(2g)^{1/2} \tan (\frac{1}{2}\Theta) h_1^{5/2}$
Truncated triangular		Broad-crested $H_1 \le 1.25 H_b$ $H_1 \ge 1.25 H_b$ $Q = C_D C_v T \frac{2}{3}(\frac{2}{3}g)^{0.5}(h_1 - \frac{1}{4}H_b)^{3/2}$ Sharp-crested $H_1 \le H_b$ $H_1 \ge H_b$ $Q = C_e \frac{8}{15}(2g)^{0.5} \dfrac{T}{H_b}[h_1^{2.5} - (h_1 - H_b)^{2.5}]$
Trapezoidal		Broad-crested $Q = C_D(T y_c + m y_c^2)[2g(H_1 - y_c)]^{1/2}$ Sharp-crested $Q = C_e \frac{2}{3}(2g)^{0.5}(b + \frac{4}{5}h_1 \tan \frac{1}{2}\Theta) h_1^{1.5}$

SOURCE: *Open-Channel Hydraulics* by R. H. French. Copyright 1985, McGraw-Hill, Inc.

TABLE 9-3 Flowing Capacities of Hydrant Nozzles, in Gallons per Minute

Pitot tube reading, psi	Inside diameter of hydrant nozzle						
	2½″	3″	4″	4½″	5″	6″	8″
1	187	269	478	604	748	1076	1912
2	264	380	676	856	1056	1520	2704
3	323	465	827	1048	1292	1860	3308
4	373	537	955	1208	1492	2148	3820
5	417	601	1068	1352	1668	2404	4272
6	457	658	1170	1480	1828	2632	4680
7	494	711	1263	1600	1976	2844	5052
8	528	760	1351	1708	2112	3040	5404
9	560	806	1433	1812	2240	3224	5732
10	590	850	1510	1912	2360	3400	6040
11	619	891	1584	2004	2476	3564	6336
12	646	931	1650	2096	2584	3724	6600
13	673	969	1722	2180	2692	3876	6888
14	698	1005	1787	2264	2792	4020	7148
15	722	1040	1849	2344	2888	4060	7396
16	746	1075	1910	2420	2984	4300	7640
17	769	1108	1969	2492	3076	4432	7876
18	791	1140	2026	2568	3164	4560	8104

TABLE 9-3 Flowing Capacities of Hydrant Nozzles, in Gallons per Minute (Continued)

Pitot tube reading, psi	Inside diameter of hydrant nozzle							
	2½″	3″	4″	4½″	5″	6″	8″
19	813	1171	2082	2636	3252	4684	8328
20	834	1201	2136	2704	3336	4804	8544
22	875	1260	2240	2836	3500	5040	8960
24	914	1316	2340	2964	3656	5264	9360
26	951	1370	2435	3084	3804	5480	9740
28	987	1422	2527	3200	3948	5688	10108
30	1022	1472	2616	3312	4086	5888	10464
32	1055	1520	2702	3424	4220	6080	10808
34	1088	1566	2785	3528	4332	6264	11140
36	1119	1612	2866	3632	4476	6448	11464
38	1150	1656	2944	3728	4600	6624	11776
40	1180	1699	3021	3824	4720	6796	12084
42	1209	1741	3095	3920	4836	6964	12380
44	1237	1782	3168	4012	4948	7128	12672
46	1265	1822	3239	4100	5060	7288	12956
48	1293	1861	3309	4188	5172	7444	13236
50	1319	1900	3377	4276	5276	7600	13508

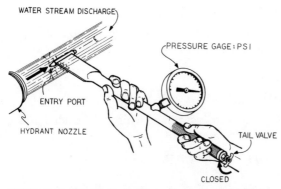

FIGURE 9-12 Pitot tube design and use. A pitot tube is a simple water meter constructed of brass or copper tubing with a standard pressure gauge threaded into the assembly. The entry port is placed into the centerline of the water stream discharging from the hydrant. Remember that the pitot tube must be held firmly during this test. After water has fully displaced the air within the tube assembly, the tail valve is closed, which activates the pressure gauge as it registers the stream's discharge pressure. Record the pressure for later use with Table 9-3. Following the test and prior to returning the pitot tube to its case, open the tail valve and drain all residual water from the assembly. Now refer to Table 9-3, using the pressure reading and the correct discharge pipe size columns, to obtain approximate gallons per minute discharge.

9-13   Flow Test Procedure

The following equipment is needed to conduct a flow test on a water distribution system:

1. Fire hydrant wrench
2. Pitot tube (see Fig. 9-12)
3. Pressure gauge (0 to 150 psi range) which is threaded into a cap for connecting to a hydrant
4. Steel ruler graduated to $\frac{1}{16}$ inch
5. Pipe wrench

The first step is to record the water system's static pressure. This is generally done by connecting the hydrant cap (with pressure gauge) to a nozzle on a fire hydrant. Using the hydrant wrench, open the hydrant very slowly and allow the air to escape through the loosely connected cap. After the residual air has bled out of the hydrant, securely tighten the cap onto the hydrant nozzle with your pipe wrench. Now fully open the hydrant with the hydrant wrench so that full system pressure can be measured by the gauge. Record this reading as "static" pressure. This force is the normal, constant system pressure when no water is being drawn through service pipes. If the gauge pointer pulsates, the reading should be taken as the center point of the swing. Leave this hydrant open and the nozzle cap gauge in place.

The second phase is to select a hydrant for the flow test. If possible, this hydrant should be located beyond the static hydrant in a direction away from the source of water supply, but within a reasonable distance. Remove the nozzle cap and measure the inside diameter to the nearest $\frac{1}{16}$ inch. Now, slowly open the hydrant via the top nut with the hydrant wrench very slowly until it is fully discharging maximum flow and stabilized.

The third step now begins. Open the tail valve on the pitot tube. Then firmly grip the main tube with both hands as shown in Fig. 9-12. Place the entry port (orifice) into the centerline of the discharging stream. After all the air in the tube has been replaced with water, close the tail valve. This will close the system and allow the pressure gauge to register the discharge force of the escaping water. At this same moment, your assistant must read the pressure at the first (static) hydrant. Both of these pressures must be recorded for a later calculation.

The fourth step is to slowly close both hydrants and replace their nozzle caps. Don't forget to open the pitot tube's tail valve to allow the trapped water to drain out. Damage could result in cold weather if exposed to freezing conditions.

The final step in this field work is performing a simple calculation to determine the flow available at a desired pressure.

Example: It is desired to determine the available water flow in the system while maintaining a 20-psi residual pressure. The following data is assumed to have been obtained by the method described previously:

Static pressure	61 psi
Residual pressure (during test)	42 psi
Nozzle inside diameter	2½ inches
Pitot tube pressure	12 psi
Discharge rate (Table 9-3)	646 gpm

Formula:
$$Q_r = Q_f \times \frac{h_r^{0.54}}{h_f^{0.54}}$$

where Q_r = available gpm with 20 psi residual
Q_f = flow (gpm) at pitot pressure
h_r = static pressure − 20 psi
h_f = static pressure − residual pressure

Therefore $$Q_r = 646 \times \frac{(61 - 20)^{0.54}}{(61 - 42)^{0.54}}$$

$$= 978 \text{ gpm at 20 psi residual}$$

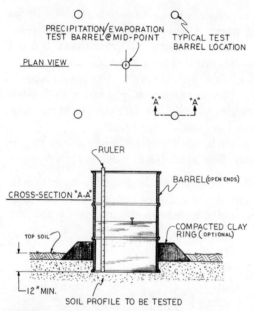

FIGURE 9-13   Barrel testing method for infiltration assessment. Preliminary on-site, field testing to determine the vertical infiltration rate of a specific area may be accomplished inexpensively by the following method.

Obtain a number of clean oil barrels and remove their lids and cut out the bottoms of all except one. Install the bottomless cylinders as shown and place the undisturbed barrel in the center of the network. A few flooding events of the bottomless barrels are necessary to ensure saturation of the soil profile to produce a near-steady state. At this point in time, all barrels should be approximately half full and the water elevations accurately recorded. Periodic measurements of the water height in all barrels must be taken throughout the test period. This monitoring data then is tabulated, taking into account the evaporation/precipitation factor from the center barrel, and reduced to an actual "infiltration versus time" curve.

It should be noted that small cylinders may be used. These cylinder infiltrometers, generally, are 6 to 14 inches in diameter and carefully driven or pushed into the soil to a depth of approximately 4 to 6 inches. Their length is recommended to be at least 12 inches. Infiltration capacity of soils often show wide variations within a given area. Therefore, these field tests should be averaged and a safety factor incorporated. In the final analysis, the designer should use additional tests, if warranted, because there is no known test or test combination that can accurately predict the actual infiltration rate of a large area prior to full-blown usage. However, this test will provide an excellent general assessment.

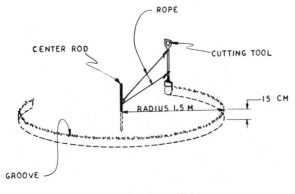

GROOVE PREPARATION

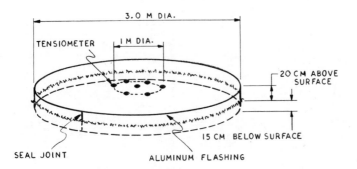

FINISHED INSTALLATION

FIGURE 9-14 Flooding basin test for infiltration rates. The U.S. Army Corps of Engineers have developed a field testing method to determine infiltration rates. This procedure is similar to the foregoing simple barrel test. This pilot-scale technique represents an excellent method for determining vertical infiltration. According to EPA's *Process Design Manual — Land Treatment of Municipal Wastewater,* about 4 worker-hours and less than 265 gallons (1000 liters) of water are required. The basics for the setup are illustrated and the basin must be flooded to saturate (or nearly so) the upper soil profile. This may be verified by the tensiometers. When the tensiometer readings approach zero through the upper soil profile, the final flooding event is monitored to record the cumulative intake (infiltration) versus time. Refer to *Simplified Field Procedures for Determining Vertical Moisture Flow Rates in Medium to Fine Textured Soils,* Engineer Technical Letter 1980, by U.S. Army Corps of Engineers.

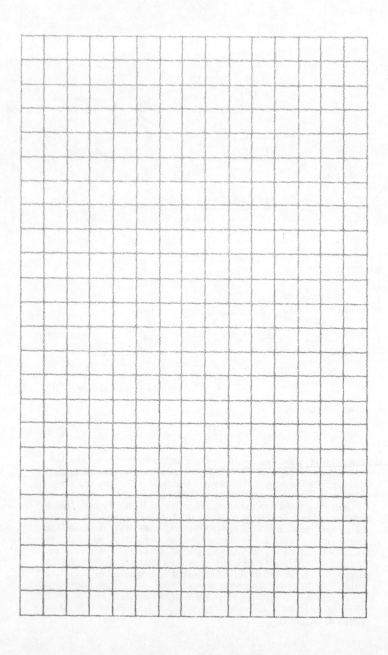

NOTES

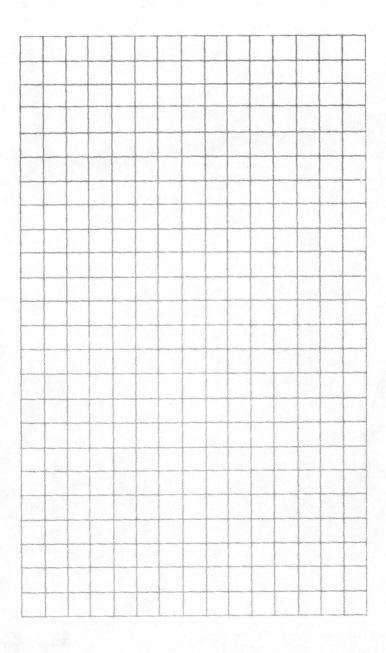

CELL #1

Supplemental Data

TABLE 10-1 Commonly Used Constants
Velocity Heads (v²/2g)

Velocity (feet per second)	Velocity head (feet)	Velocity (feet per second)	Velocity head (feet)
2.0	0.062	6.0	0.559
2.2	0.075	6.2	0.597
2.4	0.089	6.4	0.636
2.6	0.105	6.6	0.675
2.8	0.122	6.8	0.718
3.0	0.140	7.0	0.761
3.2	0.159	7.2	0.805
3.4	0.180	7.4	0.850
3.6	0.201	7.6	0.897
3.8	0.224	7.8	0.945
4.0	0.248	8.0	0.994
4.2	0.274	8.2	1.044
4.4	0.301	8.4	1.096
4.6	0.329	8.6	1.148
4.8	0.351	8.8	1.202
5.0	0.388	9.0	1.258
5.2	0.420	9.2	1.314
5.4	0.453	9.4	1.372
5.6	0.487	9.6	1.431
5.8	0.522	9.8	1.491

1 CFS	= 0.646 MGD	⅛ inch	= 0.01 feet approx.
1.55 CFS	= 1 MGD	1 cu. ft.	= 7.48 gallons
1 MGD	= 694.44 GPM	1 gal.	= 0.134 cu. ft.
1 ft. head	= 0.433 psi	1 sq. mile	= 640 acres
1 psi	= 2.31 ft. head	1 acre	= 43,560 sq. ft.

SOURCE: National Clay Pipe Institute.

10-1   Conversions, Constants, and Formulas

Volume and Weight

1 U. S. gallon = 8.34 lbs × Sp Gr
1 U. S. gallon = 0.84 Imperial gallon
1 cu ft of liquid = 7.48 gal
1 cu ft of liquid = 62.32 lbs × Sp Gr
Specific gravity of sea water = 1.025 to 1.03
1 cu meter = 264.5 gal
1 barrel (oil) = 42 gal

Capacity and Velocity

1 gpm = 449 cu ft per sec

$$\text{gpm} = \frac{\text{lbs per hour}}{500 \times \text{Sp. Gr.}}$$

gpm = 0.069 × boiler Hp
gpm = 0.7 × bbl /hour = 0.0292 bbl /day
gpm = 0.227 metric tons per hour

1 mgd = 694.5 gpm

$$V = \frac{\text{gpm} \times 0.321}{\text{area in sq in}} = \frac{\text{gpm} \times 0.409}{D^2}$$

$$V = \sqrt{2gH}$$

gpm = gallons per minute
Sp Gr = specific gravity based on water at 62°F

Hp = horsepower
bbl = barrel (oil) = 42 gal
mgd = million gallons per day of 24 hours
V = velocity in ft /sec
D = diameter in inches
g = 32.16 ft /sec /sec
H = head in feet

Head

$$\text{Head in feet} = \frac{\text{Head in psi} \times 2.31}{\text{Sp Gr}}$$

1 foot water (cold, fresh) = 1.133 inches of mercury
1 psi = 0.0703 kilograms per sq centimeter
1 psi = 0.068 atmosphere

$$H = \frac{V^2}{2g}$$

psi = pounds per square inch

Power and Torque

1 horsepower = 550 ft-lb per sec
= 33,000 ft-lb per min
= 2545 btu per hr
= 745.7 watts
= 0.7457 kilowatts

Power and Torque—(*Continued*)

$$\text{bhp} = \frac{\text{gpm} \times \text{Head in feet} \times \text{Sp Gr}}{3960 \times \text{efficiency}}$$

$$\text{bhp} = \frac{\text{gpm} \times \text{Head in psi}}{1714 \times \text{efficiency}}$$

Navy formula to determine Hp rating of motor:

$$\text{Hp} = \text{ohp} \underbrace{\left(1.05 + \frac{1.35}{\text{ohp} + 3} \right)}_{\text{efficiency of pump}}$$

where ohp is the output horsepower or water horse-power work done by the pump which is determined by:

$$\text{ohp} = \frac{\text{gpm} \times \text{Head in feet} \times \text{Sp Gr}}{3960}$$

$$\text{or ohp} = \frac{\text{gpm} \times \text{Head in psi}}{1714}$$

$$\text{Torque in lbs feet} = \frac{\text{Hp} \times 5252}{\text{rpm}}$$

bhp = brake horsepower
rpm = revolutions per minute

Miscellaneous Centrifugal Pump Formulas

$$\text{Specific speed} = N_s = \frac{\sqrt{\text{gpm}} \times \text{rpm}}{H^{3/4}}$$

where H = head per stage in feet

$$\text{Diameter of impeller in inches} = d = \frac{1840 \, K_u \sqrt{H}}{\text{rpm}}$$

where Ku is a constant varying with impeller type and design. Use H at shut-off (zero capacity) and Ku is approx. 1.0

$$\text{At constant speed:} \quad \frac{d_1}{d_2} = \frac{\text{gpm}_1}{\text{gpm}_2} = \frac{\sqrt{H_1}}{\sqrt{H_2}} = \frac{\sqrt{\text{bhp}_1}}{\sqrt{\text{bhp}_2}}$$

At constant impeller diameter

$$\frac{\text{rpm}_1}{\text{rpm}_2} = \frac{\text{gpm}_1}{\text{gpm}_2} = \frac{\sqrt{H_1}}{\sqrt{H_2}} = \frac{\sqrt{\text{Bhp}_1}}{\sqrt{\text{Bhp}_2}}$$

SOURCE: *Pump Handbook* by I. J. Karassik et al. Copyright 1976, McGraw-Hill, Inc.

10-2   Quantity and Velocity Equations

The following equations are provided to show the basis for flow diagrams and to supply equations for more accurate hydraulic calculations. The designer is reminded that precise calculations of hydraulic data are not possible except under controlled conditions.

Manning equations

The most commonly used velocity and quantity equations are

$$V = \frac{1.486}{n} \, r^{2/3} s^{1/2} \qquad \text{(velocity)}$$

$$Q = \frac{1.486}{n} \, ar^{2/3} s^{1/2} \qquad \text{(quantity)}$$

V is the velocity of flow (averaged over the cross section of the flow) measured in feet per second. For sewers flowing at design depth, V should exceed 2 feet per second to prevent settlement of solids in the pipe. Conversely, velocities exceeding 20 feet per second should be avoided where possible. Clay pipe can handle high velocities without damage; however, manholes, structures, and angle points must be designed carefully to avoid problems.

Q is the quantity of flow measured in cubic feet per second.

n is a coefficient of roughness which is used in Manning's equation to calculate flow in a pipe.

a represents the cross-sectional area of the flowing water in square feet.

r represents the hydraulic radius of the wetted cross section of the pipe measured in feet. It is obtained by dividing a by the length of the wetted perimeter.

s represents the slope of the energy gradient. It is numerically equal to the slope of the invert and the hydraulic surface in uniform flow.

This subsection courtesy of National Clay Pipe Institute.

TABLE 10-2   Values of n for the Manning Formula:
$q = (1/n)AR^{2/3}S^{1/2}$

Channel condition	n†
Plastic, glass, drawn tubing	0.009
Neat cement, smooth metal	0.010
Planed timber, asbestos pipe	0.011
Wrought iron, welded steel, canvas	0.012
Ordinary concrete, asphalted cast iron	0.013
Unplaned timber, vitrified clay, glazed brick	0.014
Cast-iron pipe, concrete pipe	0.015
Riveted steel, brick, dressed stone	0.016
Rubble masonry	0.017
Smooth earth	0.018
Firm gravel	0.020
Corrugated metal pipe and flumes	0.023
Natural channels:	
Clean, straight, full stage, no pools	0.029
As above with weeds and stones	0.035
Winding, pools and shallows, clean	0.039
As above at low stages	0.047
Winding, pools and shallows, weeds and stones	0.042
As above, shallow stages, large stones	0.052
Sluggish, weedy, with deep pools	0.065
Very weedy and sluggish	0.112

† Values quoted are averages of many determinations: variations of as much as 20 percent must be expected, especially in natural channels.

SOURCE: *Hydrology for Engineers* by R. K. Linsley et al. Copyright 1982, McGraw-Hill, Inc.

TABLE 10-3   Density and Volume of Water

Temperature (°C)	(°F)	Density (g/cc)	Volume (cc/g)
−10	14.0	0.99815	1.00186
−8	17.6	0.99869	1.00131
−6	21.2	0.99912	1.00088
−4	24.8	0.99945	1.00055
0	32.0	0.99987	1.00013
4	39.2	1.00000	1.00000
6	42.8	0.99997	1.00003
8	46.4	0.99988	1.00012
10	50.0	0.99973	1.00027
15	59.0	0.99913	1.00087
20	68.0	0.99823	1.00177
25	77.0	0.99708	1.00293
30	86.0	0.99568	1.00434
35	95.0	0.99406	1.00598
40	104.0	0.99225	1.00782
45	113.0	0.99025	1.00985
50	122.0	0.98807	1.01207
55	131.0	0.98573	1.01448
60	140.0	0.98324	1.01705
70	158.0	0.97781	1.02270
80	176.0	0.97183	1.02899
90	194.0	0.96534	1.03590
100	212.0	0.95838	1.04343
110	230.0	0.9510	1.0515
120	248.0	0.9434	1.0601
140	284.0	0.9264	1.0794
160	320.0	0.9075	1.1019
180	356.0	0.8866	1.1279
200	392.0	0.8628	1.1590

SOURCE: *Fluid Flow Pocket Handbook* by Nicholas P. Cheremisinoff. Copyright 1984, Gulf Publishing Co., Houston, Texas. Used with permission. All rights reserved.

TABLE 10-4 Conversion Units Chart for Volumetric Flow Rate

Given Units of:	Multiply by Table Values to Convert to These Units											
	m^3/s	dm^3/s	ft^3/d	ft^3/hr	ft^3/min	ft^3/s	U.K. gal/hr	U.S. gal/hr	U.K. gal/min	U.S. gal/min	bbl/d	bbl/hr
m^3/s	1	10^3	3.05119×10^6	1.2713×10^5	2.1189×10^3	3.5315×10^1	7.9189×10^5	9.5102×10^5	1.3198×10^4	1.5850×10^4	5.4344×10^5	2.2643×10^4
dm^3/s	10^{-3}	1	3.05119×10^3	1.2713×10^2	2.1189	3.5315×10^{-2}	7.9189×10^2	9.5102×10^2	1.3198×10^1	1.5850×10^1	5.4344×10^2	2.2643×10^1
ft^3/d	3.277×10^{-7}	3.277413×10^{-4}	1	4.1667×10^{-2}	6.9444×10^{-4}	1.15741×10^{-5}	3.7429×10^{-1}	3.1167×10^{-1}	6.2383×10^{-3}	5.1940×10^{-3}	1.781×10^{-1}	7.421×10^{-3}
ft^3/hr	7.866×10^{-6}	7.865791×10^{-3}	24	1	1.6667×10^{-2}	2.7778×10^{-4}	8.9831	7.48	1.4972×10^{-1}	1.2466×10^{-1}	4.274	1.781×10^{-1}
ft^3/min	4.719×10^{-4}	4.719474×10^{-1}	1.4400×10^3	60	1	1.6667×10^{-2}	5.3897×10^2	4.488×10^2	8.983	7.48	2.565×10^2	1.069×10^1
ft^3/s	2.832×10^{-2}	2.831685×10^1	8.6400×10^4	3600	60	1	3.234×10^4	2.693×10^4	5.3897×10^2	4.488×10^2	1.539×10^4	6.411×10^2
U.K. gal/hr	1.263×10^{-6}	1.262803×10^{-3}	2.6717	1.1132×10^{-1}	1.8554×10^{-3}	3.0923×10^{-5}	1	8.327×10^{-1}	1.667×10^{-2}	1.3878×10^{-2}	4.758×10^{-1}	1.983×10^{-2}
U.S. gal/hr	1.052×10^{-6}	1.051503×10^{-3}	3.20856	1.3369×10^{-1}	2.2282×10^{-3}	3.7136×10^{-5}	1.20094	1	2.00157×10^{-2}	1.667×10^{-2}	5.714×10^{-1}	2.381×10^{-2}
U.K. gal/min	7.577×10^{-5}	7.576820×10^{-2}	1.6030×10^2	6.6793	1.1132×10^{-1}	1.8554×10^{-3}	60	4.9961×10^1	1	8.3268×10^{-1}	2.855×10^1	1.189
U.S. gal/min	6.309×10^{-5}	6.309020×10^{-2}	1.9253×10^2	8.0220	1.337×10^{-1}	2.228×10^{-3}	7.2056×10^1	60	1.20094	1	3.428×10^1	1.429
bbl/d	1.840×10^{-6}	1.840131×10^{-3}	5.615	2.3396×10^{-1}	3.899×10^{-3}	6.499×10^{-5}	2.1017	1.750	3.503×10^{-2}	2.917×10^{-2}	1	4.1667×10^{-2}
bbl/hr	4.416×10^{-5}	4.416314×10^{-2}	1.3476×10^2	5.615	9.358×10^{-2}	1.5597×10^{-3}	5.044×10^1	42	8.407×10^{-1}	7.000×10^{-1}	24	1

SOURCE: *Fluid Flow Pocket Handbook* by Nicholas P. Cheremisinoff. Copyright 1984, Gulf Publishing Co., Houston, Texas. Used with permission. All rights reserved.

TABLE 10-5 Conversion Units Chart for Force

Given Units Of:	Multiply by Table Values to Convert to These Units					
	$g\text{-}cm\text{-}s^{-2}$ (dyne)	$kg\text{-}m\text{-}s^{-2}$ (N)	$lb_m\text{-}ft\text{-}s^{-2}$ (poundal)	lb_f	U.K. ton f	U.S. ton f
$g\text{-}cm\text{-}s^{-2}$ (dyne)	1	10^{-5}	7.2330×10^{-5}	2.2481×10^{-6}	1.004×10^{-3}	1.124×10^{-3}
$kg\text{-}m\text{-}s^{-2}$ (N)	10^5	1	7.2330	2.2481×10^{-1}	100.4	112.4
$lb_m\text{-}ft\text{-}s^{-2}$ (poundal)	1.3826×10^4	1.3826×10^{-1}	1	3.1081×10^{-2}	1.388×10^1	1.554×10^1
lb_f	4.4482×10^5	4.4482	32.1740	1	4.464×10^2	5.00×10^2
U.K. ton f	9.964×10^2	9.964×10^{-3}	7.207×10^{-2}	2.240×10^{-3}	1	1.120
U.S. ton f	8.896×10^2	8.896×10^{-3}	6.435×10^{-2}	2.000×10^{-3}	0.8929	1

SOURCE: *Fluid Flow Pocket Handbook* by Nicholas P. Cheremisinoff. Copyright 1984, Gulf Publishing Co., Houston, Texas. Used with permission. All rights reserved.

TABLE 10-6 Conversion Units Chart for Pressure

Given Units Of:	Multiply by Table Values to Convert to These Units										
	g/cm-s^2 (dyne/cm^2)	kg/m-s^2 (N/m^2)	lb$_m$/ft-s^2 (poundal/ft^2)	lb$_f$/ft^2	lb$_f$/in.^{2} (psi)	Atmospheres (Atm)	mm Hg	in. Hg	bar	Pa	kPa
g/cm-s^2 (dyne/cm^2)	1	10^{-1}	6.7197×10^{-2}	2.0886×10^{-3}	1.4504×10^{-5}	9.8692×10^{-7}	7.5006×10^{-4}	2.9530×10^{-5}	10^{-6}	10^{-1}	10^{-4}
kg/m-s^2 (N/m^2)	10	1	6.7197×10^{-1}	2.0886×10^{-2}	1.4504×10^{-4}	9.8692×10^{-6}	7.5006×10^{-3}	2.9530×10^{-4}	10^{-5}	1.000	1.000×10^{-3}
lb$_m$/ft-s^2 (poundal/ft^2)	1.4882×10^{1}	1.4882	1	3.1081×10^{-2}	2.1584×10^{-4}	1.4687×10^{-5}	1.1162×10^{-2}	4.3945×10^{-4}	1.488×10^{-5}	1.488	1.488×10^{-3}
lb$_f$/ft^2	4.7880×10^{2}	4.7880×10^{1}	32.1740	1	6.9444×10^{-3}	4.7254×10^{-4}	3.5913×10^{-1}	1.4139×10^{-2}	4.78803×10^{-4}	4.78803×10^{1}	4.78803×10^{-2}
lb$_f$/in.^{2}	6.8947×10^{4}	6.8947×10^{3}	4.6330×10^{3}	144	1	6.8046×10^{-2}	5.1715×10^{1}	2.0360	6.89476×10^{-2}	6.89476×10^{3}	6.89476
Atmospheres (atm)	1.0133×10^{6}	1.0133×10^{5}	6.8087×10^{4}	2.1162×10^{3}	14.696	1	760	29.921	1.01325	1.01325×10^{5}	1.01325×10^{2}
mm Hg	1.3332×10^{3}	1.3332×10^{2}	8.9588×10^{1}	2.7845	1.9337×10^{-2}	1.3158×10^{-3}	1	3.9370×10^{-2}	1.33322×10^{-3}	1.33322×10^{2}	1.33322×10^{-1}
in. Hg	3.3864×10^{4}	3.3864×10^{3}	2.2756×10^{3}	7.0727×10^{1}	4.9116×10^{-1}	3.3421×10^{-2}	25.400	1	3.38638×10^{-2}	3.38638×10^{3}	3.38638
bar	10^{6}	10^{5}	6.720×10^{4}	2.088×10^{3}	1.450×10^{1}	9.869×10^{-1}	7.5006×10^{2}	2.953×10^{1}	1	10^{5}	100
Pa	10	1	6.720×10^{-1}	2.089×10^{-2}	1.450×10^{-4}	9.869×10^{-6}	7.5006×10^{-3}	2.953×10^{-4}	10^{-5}	1	10^{-3}
kPa	10^{4}	10^{3}	6.720×10^{2}	2.089×10^{1}	1.450×10^{-1}	9.869×10^{-3}	7.5006	2.953×10^{-1}	10^{-2}	10^{3}	1

SOURCE: *Fluid Flow Pocket Handbook* by Nicholas P. Chereminisoff. Copyright 1984, Gulf Publishing Co., Houston, Texas. Used with permission. All rights reserved.

TABLE 10-7 Properties of Water in Metric Units

Temp., °C	Specific gravity	Density, g/cm³	Heat of vaporization, cal/g	Viscosity Dynamic, centipoise‡	Viscosity Kinematic, centistokes§	Vapor pressure mm Hg	Vapor pressure Millibar	Vapor pressure g/cm²
0	0.99987	0.99984	597.3	1.79	1.79	4.58	6.11	6.23
5	0.99999	0.999967†	594.5	1.52	1.52	6.54	8.72	8.89
10	0.99973	0.99970	591.7	1.31	1.31	9.20	12.27	12.51
15	0.99913	0.99910	588.9	1.14	1.14	12.78	17.04	17.38
20	0.99824	0.99821	586.0	1.00	1.00	17.53	23.37	23.83
25	0.99708	0.99705	583.2	0.890	0.893	23.76	31.67	32.30
30	0.99568	0.99565	580.4	0.798	0.801	31.83	42.43	43.27
35	0.99407	0.99404	577.6	0.719	0.723	42.18	56.24	57.34
40	0.99225	0.99222	574.7	0.653	0.658	55.34	73.78	75.23
50	0.98807	0.98804	569.0	0.547	0.554	92.56	123.40	125.83
60	0.98323	0.98320	563.2	0.466	0.474	149.46	199.26	203.19
70	0.97780	0.97777	557.4	0.404	0.413	233.79	311.69	317.84
80	0.97182	0.97179	551.4	0.355	0.365	355.28	473.67	483.01
90	0.96534	0.96531	545.3	0.315	0.326	525.89	701.13	714.95
100	0.95839	0.95836	539.1	0.282	0.294	760.00	1013.25	1033.23

† Maximum density is 0.999973 g/cm³ at 3.98°C.
‡ centipoise = (g/cm · s) × 10² = (Pa · s) × 10³
§ centistokes = (cm²/s) × 10² = (m²/s) × 10⁶

SOURCE: *Hydrology for Engineers* by R. K. Linsley et al. Copyright 1982, McGraw-Hill, Inc.

TABLE 10-8   Peak Factor

Qav = average flow, Pf = peak factor, Qpk = peak flow.

Qav	Pf	Qpk	Qav	Pf	Qpk
0 – 0.1	3.50	0 – 0.3	58 – 66	1.62	95 – 106
0.1 – 0.3	2.80	0.3 – 0.8	66 – 78	1.60	106 – 124
0.3 – 0.6	2.60	0.8 – 1.5	78 – 83	1.58	124 – 131
0.6 – 0.9	2.50	1.5 – 2.2	83 – 87	1.57	131 – 136
0.9 – 1.2	2.40	2.2 – 2.8	87 – 95	1.56	136 – 148
1.2 – 1.5	2.35	2.8 – 3.5	95 – 101	1.55	148 – 156
1.5 – 1.9	2.30	3.5 – 4.3	101 – 108	1.54	156 – 166
1.9 – 2.4	2.25	4.3 – 5.3	108 – 116	1.53	166 – 177
2.4 – 3.0	2.20	5.3 – 6.5	116 – 124	1.52	177 – 188
3.0 – 3.8	2.15	6.5 – 8.1	124 – 133	1.51	188 – 200
3.8 – 4.9	2.10	8.1 – 10.2	133 – 142	1.50	200 – 212
4.9 – 6.3	2.05	10.2 – 12.8	142 – 152	1.49	212 – 226
6.3 – 7.5	2.00	12.8 – 14.9	152 – 163	1.48	226 – 240
7.5 – 8.3	1.98	14.9 – 16.4	163 – 175	1.47	240 – 256
8.3 – 9.2	1.96	16.4 – 17.9	175 – 188	1.46	256 – 274
9.2 – 10.3	1.94	17.9 – 19.9	188 – 202	1.45	274 – 292
10.3 – 11.4	1.92	19.9 – 22.0	202 – 216	1.44	292 – 310
11.4 – 12.7	1.90	22.0 – 24.0	216 – 233	1.43	310 – 332
12.7 – 14.2	1.88	24.0 – 27.0	233 – 250	1.42	332 – 354
14.2 – 15.9	1.86	27.0 – 29.0	250 – 269	1.41	354 – 378
15.9 – 18.0	1.84	29.0 – 33.0	269 – 290	1.40	378 – 405
18.0.– 20.0	1.82	33.0 – 36.0	290 – 312	1.39	405 – 432
20.0 – 22.0	1.80	36.0 – 39.0	312 – 336	1.38	432 – 462
22.0 – 25.0	1.78	39.0 – 44.0	336 – 362	1.37	462 – 494
25.0 – 28.0	1.76	44.0 – 49.0	362 – 391	1.36	494 – 530
28.0 – 32.0	1.74	49.0 – 55.0	391 – 422	1.35	530 – 568
32.0 – 36.0	1.72	55.0 – 62.0	422 – 455	1.34	568 – 607
36.0 – 40.0	1.70	62.0 – 68.0	455 – 492	1.33	607 – 652
40.0 – 45.0	1.68	68.0 – 75.0	492 – 532	1.32	652 – 700
45.0 – 51.0	1.66	75.0 – 84.0	532 – 575	1.31	700 – 750
51.0 – 58.0	1.64	84.0 – 95.0	575 – ∞	1.30	750 – ∞

NOTES:

1. Enter table with average or peak flow. Use appropriate Peak Factor to convert to peak or average flow.
2. Peak Factors vary but generally decrease as flow increases.
3. Adjust Peak Factors when local information is available.
4. The accuracy of hydraulic calculations does not warrant interpolation of Peak Factor.

SOURCE: National Clay Pipe Institute.

R = Run-off in cubic feet per second
A = Area in acres
C = Run-off coefficient in inches of rain fall per 24 hours

$$R = \frac{121}{2880}\ AC$$

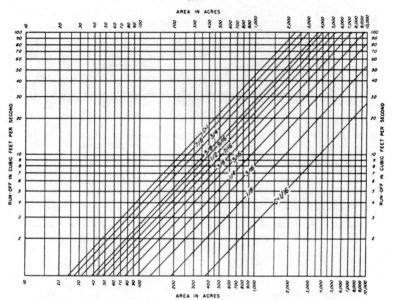

FIGURE 10-1 Surface drainage runoff diagram. *(Courtesy: National Clay Pipe Institute.)*

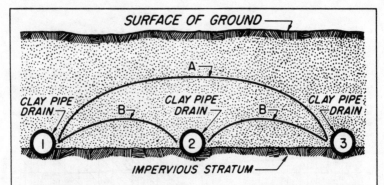

As previously shown, the amount by which ground water tables are lowered is influenced to a high degree by the spacing of the subdrains. Ground-water profiles follow the laws of hydraulic flow through porous mediums and resemble the shapes indicated in the above diagram. The ground water adjacent to an open conduit is always at the same elevation as the water in the conduit. The height to which the water rises between drains is a function of the fineness of the soil. Closer spacing between subdrains not only reduces the level of the ground water materially but also reduces, to a great degree, the amount of ground water remaining in the soil. This leaves the soil with a much greater storage capacity for subsequent percolation.

In the above diagram curve A shows the water-table for pipe lines 1 and 3 (omitting pipe 2.) Curve B shows the effect of the intermediate line. The area between curve A and B shows the increase in soil volume drained.

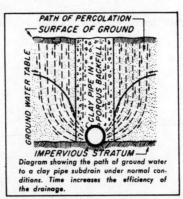

Diagram showing the path of ground water to a clay pipe subdrain under normal conditions. Time increases the efficiency of the drainage.

FIGURE 10-2   Groundwater drainage via subsurface drain pipes. *(Courtesy: National Clay Pipe Institute.)*

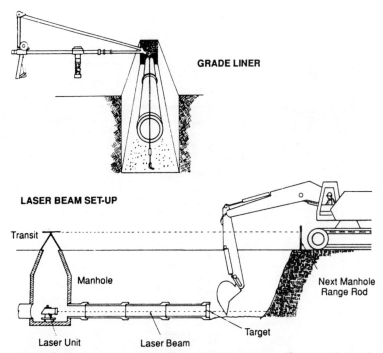

GRADE LINER

LASER BEAM SET-UP

Transit

Manhole

Next Manhole
Range Rod

Laser Unit   Laser Beam   Target

FIGURE 10-3   Laser beam setup for laying pipe to grade. *(Courtesy: National Clay Pipe Institute.)*

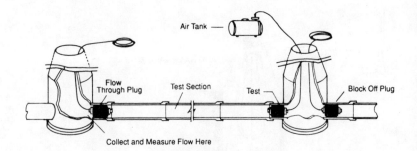

INFILTRATION TEST

Calculation of Infiltration Rate

1. Determine the rate of infiltration by collecting a known quantity of water in a recorded period of time. Example: It takes 15 minutes to collect 1 quart or 1 gallon/hour.

2. Convert to gallons/day/mile of line. Example:

$$\frac{(Gals./Hr.) \ (24 \ Hrs./Day) \ (5,280 \ Ft.)}{Line \ Length \ in \ Ft.}$$

= Gals./Mile/Day

3. Divide above flow rate by pipe diameter in inches to obtain gallons per inch of pipe diameter per mile per day.

FIGURE 10-4   Calculation of infiltration rate in pipeline. *(Courtesy: National Clay Pipe Institute.)*

TABLE 10-9   Power Consumed in Pumping 1000 Gallons of Clear Water at 1 Foot Total Head (Various Efficiencies)

Overall Efficiency Pump Unit	Kwh Per 1000 Gallons at One Ft. Total Head	Overall Efficiency Pump Unit	Kwh Per 1000 Gallons at One Ft. Total Head	Overall Efficiency Pump Unit	Kwh Per 1000 Gallons at One Ft. Total Head
32	.00980	51.5	.00609	71	.00442
32.5	.00958	52	.00603	71.5	.00439
33	.00951	52.5	.00597	72	.00435
33.5	.00937	53	.00592	72.5	.00432
34	.00922	53.5	.00586	73	.00430
34.5	.00909	54	.00581	73.5	.00427
35	.00896	54.5	.00575	74	.00424
35.5	.00884	55	.00570	74.5	.00421
36	.00871	55.5	.00565	75	.00418
36.5	.00860	56	.00560	75.5	.00415
37	.00848	56.5	.00555	76	.00413
37.5	.00837	57	.00550	76.5	.00410
38	.00826	57.5	.00545	77	.00407
38.5	.00815	58	.00541	77.5	.00405
39	.00804	58.5	.00536	78	.00402
39.5	.00794	59	.00532	78.5	.00399
40	.00784	59.5	.00527	79	.00397
40.5	.00775	60	.00523	79.5	.00394
41	.00765	60.5	.00518	80	.00392
41.5	.00756	61	.00514	80.5	.00389
42	.00747	61.5	.00510	81	.00387
42.5	.00738	62	.00506	81.5	.00385
43	.00730	62.5	.00502	82	.00382
43.5	.00721	63	.00498	82.5	.00380
44	.00713	63.5	.00494	83	.00378
44.5	.00705	64	.00490	83.5	.00375
45	.00697	64.5	.00486	84	.00373
45.5	.00689	65	.00482	84.5	.00371
46	.00682	65.5	.00479	85	.00369
46.5	.00675	66	.00475	85.5	.00367
47	.00667	66.5	.00472	86	.00365
47.5	.00660	67	.00468	86.5	.00362
48	.00653	67.5	.00465	87	.00360
48.5	.00647	68	.00461	87.5	.00358
49	.00640	68.5	.00458	88	.00356
49.5	.00634	69	.00454	88.5	.00354
50	.00627	69.5	.00451	89	.00352
50.5	.00621	70	.00448	89.5	.00350
51	.00615	70.5	.00445	90	.00348

Overall efficiency $=$ true Input-Output efficiency of motor x pump efficiency.

$Kwh/1000$ gal. $= K \cdot H$

Where $K = Kwh/1000$ gal. at one ft. head. $H =$ Total Head.

Example: Overall efficiency $= 72\%$. Total Head at the rated capacity $= 150$ ft.

$Kwh/1000$ gal. $= .00435$ x $150 = 0.653$

SOURCE: Fairbanks Morse Pump Corporation.

TABLE 10-10   Radius of Curvature and Angle of Deflection for Curvilinear Sewers Using Various Pipe Lengths

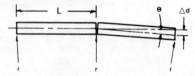

Note: Add ½ pipe O.D. to obtain centerline radius

NOMINAL PIPE Diameter (inches)	Max. Allow. Defl, △d* in./ft. of pipe (angle, Θ)	Equation for Minimum Radius of curvature, ft. (L = pipe length)	r (feet) Minimum Radius of Curvature, For Pipe Length, L, of:			
			4'	6'	8'	10'
3 to 12	½" (2.4°)	r = 24.0 (L)	96	144	192	—
15 to 24	⅜" (1.8°)	r = 32.0 (L)	128	192	256	320
27 to 36	¼" (1.2°)	r = 48.0 (L)	192	288	384	480
39 to 42	³⁄₁₆" (0.9°)	r = 64.0 (L)	256	384	512	640

*Based on ASTM C 425

SOURCE: National Clay Pipe Institute.

TABLE 10-11   Conveyance Factors (for Pipes 15 Inches and Smaller and Pipes 18 Inches and Larger)

d/D to %Qd; d/D depth over Diameter; %Qd = Q at any depth divided by Q design.

				.5D TABLE FOR PIPE 15" AND SMALLER						
			For pipe 15" and smaller, Qd = Q at a depth of .5 Diameter							
d/D	.00	.01	.02	.03	.04	.05	.06	.07	.08	.09
.0	%	0	0	0	1	1	1	2	3	3
.1	4	5	6	7	8	10	11	13	14	16
.2	18	19	21	23	25	27	30	32	34	37
.3	39	42	44	47	50	52	55	58	61	64
.4	67	70	73	77	80	83	86	90	93	96
.5	100	103	106	110	113	117	120	124	127	131
.6	134	138	141	144	148	151	154	158	161	164
.7	167	170	173	176	179	182	185	188	190	193
.8	195	197	200	202	204	206	207	209	210	212
.9	213	214	214	215	215	215	214	213	211	208
1.0	200									

Example: The depth of flow in an 8" sewer was measured at .21' d/D = .21/.67= .31. Enter table for smaller sewers with d/D = .31 and read 42% Q design. Q design is read from Design Capacity Charts.

				.75D TABLE FOR PIPE 18" AND LARGER						
			For pipe 18" and larger, Qd = Q at a depth of .75 Diameter							
d/D	.00	.01	.02	.03	.04	.05	.06	.07	.08	.09
.0	%	0	0	0	0	1	1	1	1	2
.1	2	3	3	4	5	5	6	7	8	9
.2	10	11	12	13	14	15	16	17	19	20
.3	21	23	24	26	27	29	30	32	34	35
.4	37	39	40	42	44	46	48	49	51	53
.5	55	57	59	60	62	64	66	68	70	72
.6	74	76	77	79	81	83	85	87	88	90
.7	92	94	95	97	99	100	102	103	105	106
.8	107	109	110	111	112	113	114	115	116	116
.9	117	118	118	118	118	118	118	117	116	114
1.0	110									

Example: The depth of flow in an 18" sewer was measured at 1.02' d/D = 1.02/1.5= .68. Enter table with d/D = .68 and read 88% of Q design. Q design is read from Design Capacity Charts.

SOURCE: National Clay Pipe Institute.

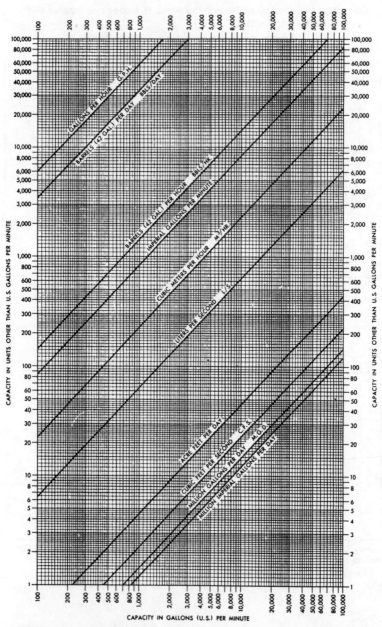

FIGURE 10-5 Capacity conversions. *(Courtesy: Pump Handbook by I. J. Karassik et al., copyright 1976, McGraw-Hill, Inc.)*

Various units	U.S. gpm
1 second-foot or cubic foot per second (cfs)	448.8
1,000,000 U.S. gallons per day (mgd)	694.4
1 imperial gallon per minute	1.201
1,000,000 imperial gallons per day	834.0
1 barrel (42 U.S. gal) per day (bbl/day)	0.0292
1 barrel per hour (bbl/hr)	0.700
1 acre-foot per day	226.3
1,000 pounds per hour (lb/hr)	2.00[1]
1 cubic meter per hour (m^3/hr)	4.403
1 liter per second (l/s)	15.851
1 metric ton per hour	4.403[1]
1,000,000 liters per day = 1,000 cubic meters per day	183.5

[1] These equivalents are based on a specific gravity of 1 for water at 62°F for English units and a specific gravity of 1 for water at 15°C for metric units. They can be used with little error for cold water of any temperature between 32°F and 80°F.

FIGURE 10-5 *(Continued)*

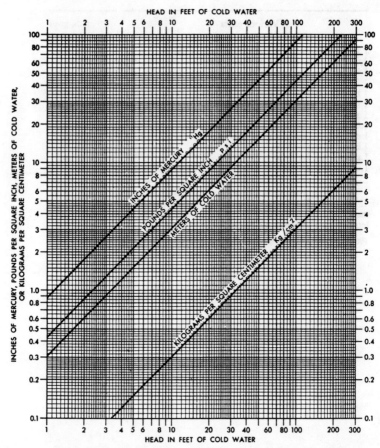

Pressure and head conversion chart. Values are plotted for 62°F (18.7°C) but can be used for water between 32° and 80°F. For liquids other than cold water, divide the head by the specific gravity (62°F water = 1.0) of the liquid at the pumping temperature to get the head in feet.

FIGURE 10-6 Pressure and head conversion chart. *(Courtesy: Pump Handbook by I. J. Karassik et al., copyright 1976, McGraw-Hill, Inc.)*

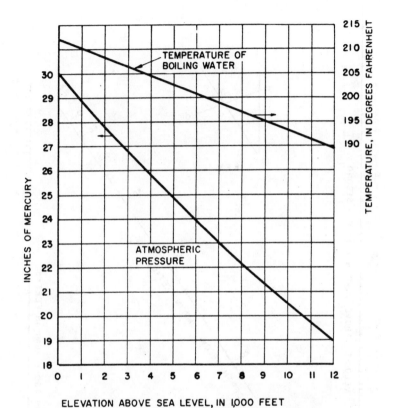

ELEVATION ABOVE SEA LEVEL, IN 1,000 FEET

FIGURE 10-7   Atmospheric pressures for altitudes up to 12,000 feet. *(Courtesy: Pump Handbook by I. J. Karassik et al., copyright 1976, McGraw-Hill, Inc.)*

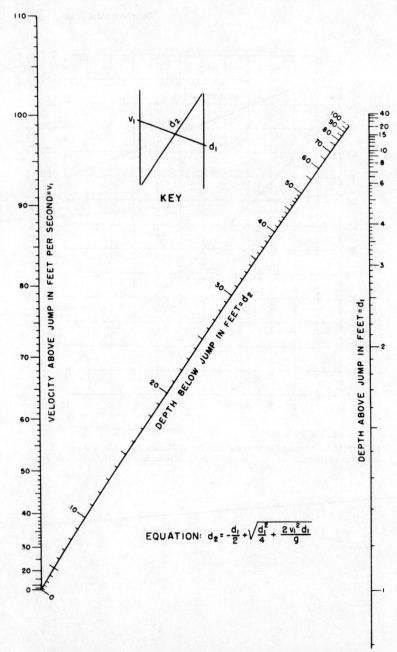

EQUATION: $d_2 = -\dfrac{d_1}{2} + \sqrt{\dfrac{d_1^2}{4} + \dfrac{2 v_1^2 d_1}{g}}$

FIGURE 10-8 Relation between variables in the hydraulic jump. *(Source: Design of Small Dams, U.S. Department of the Interior, Bureau of Reclamation.)*

316

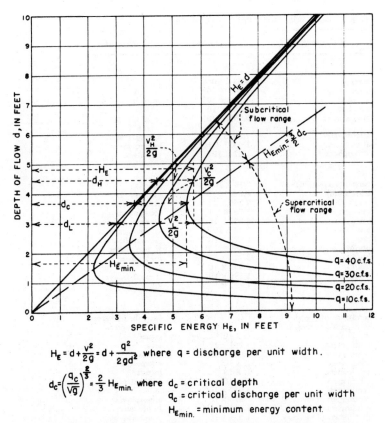

$$H_E = d + \frac{v^2}{2g} = d + \frac{q^2}{2gd^2} \quad \text{where } q = \text{discharge per unit width.}$$

$$d_c = \left(\frac{q_c}{\sqrt{g}}\right)^{\frac{2}{3}} = \frac{2}{3} H_{E_{min.}} \quad \text{where } d_c = \text{critical depth}$$
$$q_c = \text{critical discharge per unit width}$$
$$H_{E_{min.}} = \text{minimum energy content.}$$

FIGURE 10-9 Depth of flow and specific energy for rectangular section in open channel. *(Source: Design of Small Dams, U.S. Department of the Interior, Bureau of Reclamation.)*

TABLE 10-12   Measurement Conversions

To convert	Multiply by	To obtain
A		
acres	4.35×10^4	sq. ft.
acres	4.047×10^3	sq. meters
acre-feet	4.356×10^4	cu. feet
acre-feet	3.259×10^5	gallons
atmospheres	2.992×10^1	in. of mercury (at 0°C.)
atmospheres	1.0333	kgs./sq. cm.
atmospheres	1.0333×10^4	kgs./sq. meter
atmospheres	1.47×10^1	pounds/sq. in.
B		
barrels (u.s., liquid)	3.15×10^1	gallons
barrels (oil)	4.2×10^1	gallons (oil)
bars	9.869×10^{-1}	atmospheres
btu	7.7816×10^2	foot-pounds
btu	3.927×10^{-4}	horsepower-hours
btu	2.52×10^{-1}	kilogram-calories
btu	2.928×10^{-4}	kilowatt-hours
btu/hr.	2.162×10^{-1}	ft. pounds/sec.
btu/hr.	3.929×10^{-4}	horsepower
btu/hr.	2.931×10^{-1}	watts
btu/min.	1.296×10^1	ft.-pounds/sec.
btu/min.	1.757×10^{-2}	kilowatts
C		
centigrade (degrees)	$(°C \times \frac{9}{5}) + 32$	fahrenheit (degrees)
centigrade (degrees)	$°C + 273.18$	kelvin (degrees)
centigrams	$1. \times 10^{-2}$	grams
centimeters	3.281×10^{-2}	feet
centimeters	3.937×10^{-1}	inches
centimeters	$1. \times 10^{-5}$	kilometers
centimeters	$1. \times 10^{-2}$	meters
centimeters	$1. \times 10^1$	millimeters
centimeters	3.937×10^2	mils
centimeters of mercury	1.316×10^{-2}	atmospheres
centimeters of mercury	4.461×10^{-1}	ft. of water
centimeters of mercury	1.934×10^{-1}	pounds/sq. in.
centimeters/sec.	1.969	feet/min.
centimeters/sec.	3.281×10^{-2}	feet/sec.
centimeters/sec.	6.0×10^{-1}	meters/min.
centimeters/sec./sec.	3.281×10^{-2}	ft./sec./sec.
cubic centimeters	3.531×10^{-5}	cubic ft.
cubic centimeters	6.102×10^{-2}	cubic in.
cubic centimeters	1.0×10^{-6}	cubic meters
cubic centimeters	2.642×10^{-4}	gallons (u.s. liquid)
cubic centimeters	2.113×10^{-3}	pints (u.s. liquid)
cubic centimeters	1.057×10^{-3}	quarts (u.s. liquid)
cubic feet	2.8320×10^4	cu. cms.
cubic feet	1.728×10^3	cu. inches
cubic feet	2.832×10^{-2}	cu. meters
cubic feet	7.48052	gallons (u.s. liquid)
cubic feet	5.984×10^1	pints (u.s. liquid)
cubic feet	2.992×10^1	quarts (u.s. liquid)
cubic feet/min.	4.72×10^2	cu. cms./sec.
cubic feet/min.	1.247×10^{-1}	gallons/sec.
cubic feet/min.	4.720×10^{-1}	liters/sec.
cubic feet/min.	6.243×10^1	pounds water/min.
cubic feet/sec.	6.46317×10^{-1}	million gals./day
cubic feet/sec.	4.48831×10^2	gallons/min.
cubic inches	5.787×10^{-4}	cu. ft.
cubic inches	1.639×10^{-5}	cu. meters

TABLE 10-12 Measurement Conversions *(Continued)*

To convert	Multiply by	To obtain
cubic inches	2.143×10^{-5}	cu. yards
cubic inches	4.329×10^{-3}	gallons

D

degrees (angle)	1.745×10^{-2}	radians
degrees (angle)	3.6×10^{3}	seconds
degrees/sec.	2.778×10^{-3}	revolutions/sec.
dynes/sq. cm.	4.015×10^{-4}	in. of water (at 4°C.)
dynes	1.020×10^{-6}	kilograms
dynes	2.248×10^{-6}	pounds

F

fathoms	1.8288	meters
fathoms	6.0	feet
feet	3.048×10^{1}	centimeters
feet	3.048×10^{-1}	meters
feet of water	2.95×10^{-2}	atmospheres
feet of water	3.048×10^{-2}	kgs./sq. cm.
feet of water	6.243×10^{1}	pounds/sq. ft.
feet/min	5.080×10^{-1}	cms./sec.
feet/min.	1.667×10^{-2}	feet/sec.
feet/min.	3.048×10^{-1}	meters/min.
feet/min.	1.136×10^{-2}	miles/hr.
feet/sec.	1.829×10^{1}	meters/min.
feet/100 feet	1.0	per cent grade
foot-pounds	1.286×10^{-3}	btu
foot-pounds	1.356×10^{7}	ergs
foot-pounds	3.766×10^{-7}	kilowatt-hrs.
foot-pounds/min.	1.286×10^{-3}	btu/min.
foot-pounds/min.	3.030×10^{-5}	horsepower
foot-pounds/min.	3.241×10^{-4}	kg.-calories/min.
foot-pounds/sec.	4.6263	btu/hr.
foot-pounds/sec.	7.717×10^{-2}	btu/min.
foot-pounds/sec.	1.818×10^{-3}	horsepower
foot-pouhds/sec.	1.356×10^{-3}	kilowatts
furlongs	1.25×10^{-1}	miles (u.s.)

G

gallons	3.785×10^{3}	cu. cms.
gallons	1.337×10^{-1}	cu. feet
gallons	2.31×10^{2}	cu. inches
gallons	3.785×10^{-3}	cu. meters
gallons	4.951×10^{-3}	cu. yards
gallons	3.785	liters
gallons (liq. br. imp.)	1.20095	gallons (u.s. liquid)
gallons (u.s.)	8.3267×10^{-1}	gallons (imp.)
gallons of water	8.337	pounds of water
gallons/min.	2.228×10^{-3}	cu. feet/sec.
gallons/min.	6.308×10^{-2}	liters/sec.
gallons/min.	8.0208	cu. feet/hr.

H

horsepower	4.244×10^{1}	btu/min.
horsepower	3.3×10^{4}	foot-lbs./min.
horsepower	5.50×10^{2}	foot-lbs./sec.
horsepower (metric)	9.863×10^{-1}	horsepower
horsepower	1.014	horsepower (metric)
horsepower	7.457×10^{-1}	kilowatts
horsepower	7.457×10^{2}	watts
horsepower (boiler)	3.352×10^{4}	btu/hr.
horsepower (boiler)	9.803	kilowatts
horsepower-hours	2.547×10^{3}	btu
horsepower-hours	1.98×10^{6}	foot-lbs.

TABLE 10-12   Measurement Conversions *(Continued)*

To convert	Multiply by	To obtain
horsepower-hours	6.4119×10^5	gram-calories
hours	5.952×10^{-3}	weeks
	I	
inches	2.540	centimeters
inches	2.540×10^{-2}	meters
inches	1.578×10^{-5}	miles
inches	2.54×10^1	millimeters
inches	1.0×10^3	mils
inches	2.778×10^{-2}	yards
inches of mercury	3.342×10^{-2}	atmospheres
inches of mercury	1.133	feet of water
inches of mercury	3.453×10^{-2}	kgs./sq. cm.
inches of mercury	3.453×10^2	kgs./sq. meter
inches of mercury	7.073×10^1	pounds/sq. ft.
inches of mercury	4.912×10^{-1}	pounds/sq. in.
in. of water (at 4°C.)	7.355×10^{-2}	inches of mercury
in. of water (at 4°C.)	2.54×10^{-3}	kgs./sq. cm.
in. of water (at 4°C.)	5.204	pounds/sq. ft.
in. of water (at 4°C.)	3.613×10^{-2}	pounds/sq. in.
	J	
joules	9.486×10^{-4}	btu
joules/cm.	1.0×10^7	dynes
joules/cm.	1.0×10^2	joules/meter (newtons)
joules/cm.	2.248×10^1	pounds
	K	
kilograms	9.80665×10^5	dynes
kilograms	1.0×10^3	grams
kilograms	2.2046	pounds
kilograms	9.842×10^{-4}	tons (long)
kilograms	1.102×10^{-3}	tons (short)
kilograms/sq. cm.	9.678×10^{-1}	atmospheres
kilograms/sq. cm.	3.281×10^1	feet of water
kilograms/sq. cm.	2.896×10^1	inches of mercury
kilograms/sq. cm.	1.422×10^1	pounds/sq. in.
kilometers	1.0×10^5	centimeters
kilometers	3.281×10^3	feet
kilometers	3.937×10^4	inches
kilometers	1.0×10^3	meters
kilometers	6.214×10^{-1}	miles (statute)
kilometers	5.396×10^{-1}	miles (nautical)
kilometers	1.0×10^6	millimeters
kilowatts	5.692×10^1	btu/min.
kilowatts	4.426×10^4	foot-lbs./min.
kilowatts	7.376×10^2	foot-lbs./sec.
kilowatts	1.341	horsepower
kilowatts	1.434×10^1	kg.-calories/min.
kilowatts	1.0×10^3	watts
kilowatt-hrs.	3.413×10^3	btu
kilowatt-hrs.	2.655×10^6	foot-lbs.
kilowatt-hrs.	8.5985×10^3	gram calories
kilowatt-hrs.	1.341	horsepower-hours
kilowatt-hrs.	3.6×10^6	joules
kilowatt-hrs.	8.605×10^2	kg.-calories
kilowatt-hrs.	8.5985×10^3	kg.-meters
kilowatt-hrs.	2.275×10^1	pounds of water raised from 62° to 212°F.

TABLE 10-12 Measurement Conversions *(Continued)*

To convert	*Multiply by*	*To obtain*
L		
links (engineers)	1.2×10^1	inches
links (surveyors)	7.92	inches
liters	1.0×10^3	cu. cm.
liters	6.102×10^1	cu. inches
liters	1.0×10^{-3}	cu. meters
liters	2.642×10^{-1}	gallons (u.s. liquid)
liters	2.113	pints (u.s. liquid)
liters	1.057	quarts (u.s. liquid)
M		
meters	1.0×10^2	centimeters
meters	3.281	feet
meters	3.937×10^1	inches
meters	1.0×10^{-3}	kilometers
meters	5.396×10^{-4}	miles (nautical)
meters	6.214×10^{-4}	miles (statute)
meters	1.0×10^3	millimeters
meters/min.	1.667	cms./sec.
meters/min.	3.281	feet/min.
meters/min.	5.468×10^{-2}	feet/sec.
meters/min.	6.0×10^{-2}	kms./hr.
meters/min.	3.238×10^{-2}	knots
meters/min.	3.728×10^{-2}	miles/hr.
meters/sec.	1.968×10^2	feet/min.
meters/sec.	3.281	feet/sec.
meters/sec.	3.6	kilometers/hr.
meters/sec.	6.0×10^{-2}	kilometers/min.
meters/sec.	2.237	miles/hr.
meters/sec.	3.728×10^{-2}	miles/min.
miles (nautical)	6.076×10^3	feet
miles (statute)	5.280×10^3	feet
miles/hr.	8.8×10^1	ft./min.
millimeters	1.0×10^{-1}	centimeters
millimeters	3.281×10^{-3}	feet
millimeters	3.937×10^{-2}	inches
millimeters	1.0×10^{-1}	meters
minutes (time)	9.9206×10^{-5}	weeks
O		
ounces	2.8349×10^1	grams
ounces	6.25×10^{-2}	pounds
ounces (fluid)	1.805	cu. inches
ounces (fluid)	2.957×10^{-2}	liters
P		
parts/million	5.84×10^{-2}	grains/u.s. gal.
parts/million	7.016×10^{-2}	grains/imp. gal.
parts/million	8.345	pounds/million gal.
pints (liquid)	4.732×10^2	cubic cms.
pints (liquid)	1.671×10^{-2}	cubic ft.
pints (liquid)	2.887×10^1	cubic inches
pints (liquid)	4.732×10^{-4}	cubic meters
pints (liquid)	1.25×10^{-1}	gallons
pints (liquid)	4.732×10^{-1}	liters
pints (liquid)	5.0×10^{-1}	quarts (liquid)
pounds	2.56×10^2	drams
pounds	4.448×10^5	dynes
pounds	7.0×10^1	grains
pounds	4.5359×10^2	grams
pounds	4.536×10^{-1}	kilograms
pounds	1.6×10^1	ounces

TABLE 10-12 Measurement Conversions *(Continued)*

To convert	*Multiply by*	*To obtain*
pounds	3.217×10^1	poundals
pounds	1.21528	pounds (troy)
pounds of water	1.602×10^{-2}	cu. ft.
pounds of water	2.768×10^1	cu. inches
pounds of water	1.198×10^{-1}	gallons
pounds of water/min.	2.670×10^{-4}	cu. ft./sec.
pound-feet	1.356×10^7	cm.-dynes
pound-feet	1.3825×10^4	cm.-grams
pound-feet	1.383×10^{-1}	meter-kgs.
pounds/cu. ft.	1.602×10^{-2}	grams/cu. cm.
pounds/cu. ft.	5.787×10^{-4}	pounds/cu. inches
pounds/sq. in.	6.804×10^{-2}	atmospheres
pounds/sq. in.	2.307	feet of water
pounds/sq. in.	2.036	inches of mercury
pounds/sq. in.	7.031×10^2	kgs./sq. meter
pounds/sq. in.	1.44×10^2	pounds/sq. ft.

Q

quarts (dry)	6.72×10^1	cu. inches
quarts (liquid)	9.464×10^2	cu. cms.
quarts (liquid)	3.542×10^{-2}	cu. ft.
quarts (liquid)	5.775×10^1	cu. inches
quarts (liquid)	2.5×10^{-1}	gallons

R

revolutions	3.60×10^2	degrees
revolutions	4.0	quadrants
rods (surveyors' meas.)	5.5	yards
rods	1.65×10^1	feet
rods	1.98×10^2	inches
rods	3.125×10^{-3}	miles

S

slugs	3.217×10^1	pounds
square centimeters	1.076×10^{-3}	sq. feet
square centimeters	1.550×10^{-1}	sq. inches
square centimeters	1.0×10^{-4}	sq. meters
square centimeters	1.0×10^2	sq. millimeters
square feet	2.296×10^{-5}	acres
square feet	9.29×10^2	sq. cms.
square feet	1.44×10^2	sq. inches
square feet	9.29×10^{-2}	sq. meters
square feet	3.587×10^{-8}	sq. miles
square inches	6.944×10^{-3}	sq. ft.
square inches	6.452×10^2	sq. millimeters
square miles	6.40×10^2	acres
square miles	2.788×10^7	sq. ft.
square yards	2.066×10^{-4}	acres
square yards	8.361×10^3	sq. cms.
square yards	9.0	sq. ft.
square yards	1.296×10^3	sq. inches

T

temperature (°C.) +273	1.0	absolute temperature (°K.)
temperature (°C.) +17.78	1.8	temperature (°F.)
temperature (°F.) +460	1.0	absolute temperature (°R.)
temperature (°F.) −32	⁵⁄₉	temperature (°C.)
tons (long)	2.24×10^3	pounds

TABLE 10-12   Measurement Conversions *(Continued)*

To convert	Multiply by	To obtain
tons (long)	1.12	tons (short)
tons (metric)	2.205×10^3	pounds
tons (short)	2.0×10^3	pounds
W		
watts	3.4129	btu/hr.
watts	5.688×10^{-2}	btu/min.
watts	4.427×10^1	ft.-lbs./min.
watts	7.378×10^{-1}	ft.-lbs./sec
watts	1.341×10^{-3}	horsepower
watts	1.36×10^{-3}	horsepower (metric)
watts	1.0×10^{-3}	kilowatts
watt-hours	3.413	btu
watt-hours	2.656×10^3	foot-lbs.
watt-hours	1.341×10^{-3}	horsepower-hours
watt (international)	1.000165	watt (absolute)
weeks	1.68×10^2	hours
weeks	1.008×10^4	minutes
weeks	6.048×10^5	seconds

SOURCE: *Pump Handbook* by I. J. Karassik et al. Copyright 1976, McGraw-Hill, Inc.

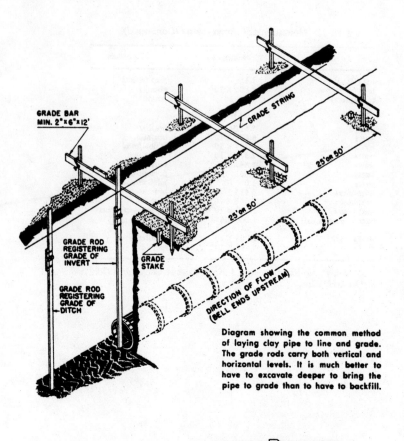

GRADE BAR
MIN. 2"x 6"x 12'

GRADE STRING

25'or 50'

25'or 50'

GRADE ROD
REGISTERING
GRADE OF
INVERT

GRADE
STAKE

GRADE ROD
REGISTERING
GRADE OF
DITCH

DIRECTION OF FLOW
(BELL ENDS UPSTREAM)

Diagram showing the common method
of laying clay pipe to line and grade.
The grade rods carry both vertical and
horizontal levels. It is much better to
have to excavate deeper to bring the
pipe to grade than to have to backfill.

The lower 90 degree arc of the
barrel of the pipe should be
in firm contact with undisturbed
earth.

Small excavations should be made
for the bells. These should be no
larger than necessary to clear
the bell.

**FIGURE 10-10   Diagram showing common method of laying pipe to line and
grade.** *(Courtesy: National Clay Pipe Institute.)*

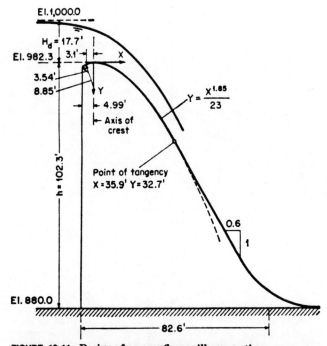

FIGURE 10-11 Design of an overflow spillway section.

Example: Determine the crest elevation and the shape of an overflow spillway section having a vertical upstream face and a crest length of 250 feet. The design discharge is 75,000 cubic feet per second. The upstream water surface at design discharge is at elevation 1000.0 and the average channel floor is at elevation 880.0 (see figure).

Solution. Assuming a high overflow spillway, the effect of approach velocity is negligible, and $C_d = 4.03$. By the discharge equation, $H_e^{1.5} = Q/CL = 75,000/(4.03 \times 250) = 74.4$ and $H_e = 17.8$ feet. The approach velocity is $V_a = 75,000/(250 \times 120) = 2.5$ feet per second, and the corresponding velocity head is $H_a = 2.5^2/2g = 0.1$ feet. Thus, the design head is $H_d = 17.8 - 0.1 = 17.7$ feet, and the height of the dam is $h = 120 - 17.7 = 102.3$ feet. This height is greater than $1.33H_d$, and, hence, the effect of approach velocity is negligible. The crest elevation is at $1000.0 - 17.7 = 982.3$. The crest shape is expressed by $Y = X^{1.85}/23$. Coordinates of the shape computed by this equation are plotted as shown in the figure.

(Courtesy: Open-Channel Hydraulics by V. T. Chow, copyright 1959, McGraw-Hill, Inc.)

NOTES

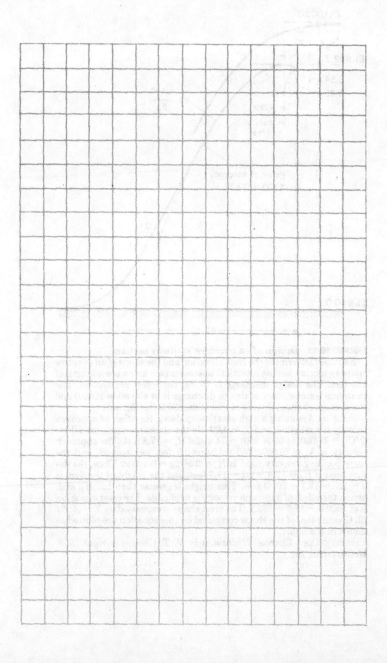

Index

ABOUT THE EDITOR-IN-CHIEF

Robert O. Parmley, P.E., CMfgE, CSI, is President and Principal Consulting Engineer of Morgan & Parmley, Ltd., Professional Consulting Engineers, Ladysmith, Wisconsin. He is also a member of the National Society of Professional Engineers, the American Society of Mechanical Engineers, the Construction Specifications Institute, the American Design Drafting Association, and the Society of Manufacturing Engineers, and is listed in the AAES *Who's Who in Engineering.* Mr. Parmley holds a BSME and a MSCE from Columbia Pacific University and is a registered professional engineer in Wisconsin, California, and Canada. He is also a certified manufacturing engineer under SME's national certification program and a certified wastewater plant operator. In a career covering more than three decades, Mr. Parmley has worked on the design and construction supervision of a wide variety of structures, systems, and machines — from dams and bridges to municipal sewage facilities and water projects. The author of over 40 technical articles, he is also the Editor-in-Chief of the *Standard Handbook of Fastening & Joining,* Second Edition, the *HVAC Field Manual,* the *Field Engineer's Manual,* and the *Mechanical Components Handbook,* all published by McGraw-Hill.